Atlas of the carabid beetles of The Netherlands

Koninklijke Nederlandse Akademie van Wetenschappen
Verhandelingen Afdeling Natuurkunde, Tweede Reeks, deel 68

Atlas of the carabid beetles of The Netherlands

H. Turin, J. Haeck and R. Hengeveld

with a computer programme by J. A. M. Snoek

North-Holland Publishing Company, Amsterdam, 1977

Author's address: Instituut voor Oecologisch Onderzoek,
Kemperbergerweg 67, Arnhem, The Netherlands

Aangeboden 27 sept. 1975, aanvaard 31 jan. 1976, gepubliceerd jan. 1977

ISBN 0 7204 8326 3

Contents

We wish to dedicate this book to C. H. Lindroth because of the enormous amount of work he has done with such great originality on so many aspects of carabid beetles.

1. Introduction

The maps presented in this Atlas summarize the present knowledge about the occurrence of carabid species in The Netherlands. The data are taken from the literature and from private and museum collections.

The resulting atlas contains, besides data on the distribution of all the carabid species occurring in The Netherlands, data on three biological characteristics: wing form and seasonal periodicity of the adults and larvae. Because the information might be useful, we have included a map showing the European distribution.

In this form, our atlas is an extension of Brakman's (1966) list of the carabids in The Netherlands, which was restricted to data on the presence or absence of each species in the eleven provinces.

During the preparatory work the first announcement was made of the European Invertebrate Survey (Heath, 1971), a publication intended to stimulate and coordinate the same type of project. Although our approach is different, this atlas may to some extent be considered one of the results of these activities.

Recently (1976), a manual prepared by the Dutch division of the E.I.S. was published giving many technical details for future participatants this project.

REFERENCES

Brakman, P. J., Lijst van Coleoptera uit Nederland en het omliggend gebied. Monographieën van de Ned. Ent. Ver. 2, Amsterdam, 1966.

Heath, J., The European Invertebrate Survey. Acta ent. fenn. 28, 27–29, 1971.

Handleiding en Atlas voor het medewerken aan de European Invertebrate Survey. Rijksmuseum v. Natuurlijke Historie, Leiden, 1976.

2. Nomenclature and identification

During the collection of the data it soon became clear that not all of the source material was of equal quality. Some of the data from the literature had to be discarded because the identification could not be checked. Inclusion of data from the literature was therefore restricted to material reported by experienced Dutch carabidologists.

On the other hand, uncertain data could usually be checked in the still existing collections. Errors in nomenclature could be traced to the original author or to changes made in the classification during or since the period covered by a particular collection. Changes in classification were followed wherever possible, although difficulties remained here too. In the latter cases, for instance *Harpalus picipennis* versus *H. vernalis*, only reliable identifications were used and doubtful ones were dropped. Changes in nomenclature were taken from the literature cited at the end of this section; the numbers there refer to Table 1.

Table 1. List of 'difficult species' (relevant literature references at the end of section 2)

no.	species name		reference nos.	no.	species name		reference nos.
33	*Nebria*	*brevicollis*	4,16	177	*Harpalus*	*puncticeps*	9
34		*salina*	4,16	191		*winkleri*	1
69	*Miscodera*	*arctica*	5	192		*latus*	1
91	*Bembidion*	*monticola*	9	193		*luteicornis*	1
92		*nitidulum*	9	201		*picipennis*	9
93		*stephensi*	9	202		*vernalis*	9
94		*milleri*	9	241	*Amara*	*kulti*	2
96		*rupestre*	4	246		*communis*	4
97		*concinnum*	4	247		*convexior*	4
98		*ustulatum*	4	248		*pseudocommunis*	4
99		*femoratum*	4	259		*ingenua*	2
158	*Badister*	*bipustulatus*	4	267		*apricaria*	3
159		*lacertosus*	4	269		*majuscula*	3
169	*Harpalus*	*rupicola*	9	270		*consularis*	3
170		*brevicollis*	9	329	*Agonum*	*micans*	17
172		*cordatus*	9	330		*scitulum*	17
174		*puncticollis*	9	332		*gracile*	17
175		*melleti*	9	337		*krynickii*	3
176		*zigzag*	5				

Despite these precautions, some of the maps still have points which deviate too much from what might be expected from the position of the other points. These deviating points have been checked whenever possible in the collections. In the few instances where this could not be done, to remain on the safe side the point was discarded.

The nomenclature used on the maps follows Brakman (1966), from whom we also adopted the species numbers. During the preparation of the maps, Lindroth's 'Handbooks for the identification of British Insects: Carabidae' (1974) appeared, in which the author used a new nomenclature for a number of species. The new names are indicated in italics on the maps.

REFERENCES

1. Boer, P. J. den, Een verwarrende onjuistheid in de 'Coleoptera Neerlandica', I. Ent. Ber., Amst. 21 (2), 41–43, 1961.

2. Boer, P. J. den, Twee nieuwe Amara soorten voor de Nederlandse fauna. Ent. Ber., Amst. 21 (8), 147–152, 1961.

3. Boer, P. J. den, Twee nieuwe loopkeversoorten voor de Nederlandse fauna. Ent. Ber., Amst. 22 (5), 88–95, 1962.

4. Brakman, P. J., Korte coleopterologische notities IV. Ent. Ber., Amst. 21 (1), 8–14, 1961.

5. Brakman, P. J., Korte coleopterologische notities VI. Ent. Ber., Amst. 23, 202–203, 1963.

6. Dahl, T. (Mrozek-), Coleoptera oder Käfer. I: Carabidae (Laufkäfer). Dahl: Tierwelt Deutschlands etc. 7, Jena, 1928.

7. Everts, E., Coleoptera Neerlandica I, II and III. Den Haag, 1898, 1903 and 1922.

8. Jeannel, R., Coléoptères Carabiques 1, 2. Faune de France 39, 40, Paris, 1941.

9. Klynstra, B. H., Mededelingen over de Nederlandse Adephaga I. Ent. Ber., Amst. 10 (1), 97–108, 1939.

10. Klynstra, B. H., Het genus Notiophilus in Nederland. Ent. Ber., Amst. 14 (1), 51–54, 1952.

11. Klynstra, B. H., Het genus Dyschirius in Nederland. Ent. Ber., Amst. 15 (11), 233–238, 1954.

12. Klynstra, B. H., Het genus Dyschirius in Nederland. Ent. Ber., Amst. 15 (12), 263–269, 1954.

13. Kooi, R. E. and Th. v. Egmond, Determinatieschema voor het genus Amara (Intern rapport) Leiden, 1972.

14. Kuhnt, P., Illustrierte Bestimmungstabellen der Käfer Deutschlands. Stuttgart, 1913.

15. Kult, K., The Carabids from Czechoslovakya. Praag, 1947.

16. Land, J. v.d., Nebria brevicollis and allied species in Western Europe (Col., Car.) Ent. Ber., Amst. 24 (1–3), 1964.

17. Lindroth, C. H., Handbooks for the identification of British Insects, Coleoptera: Carabidae. London, 1974.

18. Reitter, E., Fauna Germanica. Die Käfer des Deutschen Reiches I. Stuttgart, 1908.

19. Wiel, P. v.d., Bijdrage tot de kennis van de Nederlandse kevers IV. Tijdschr. Ent. 99 (1–2), 2–3, 1956.

20. Wiel, P. v.d., Bijdrage tot de kennis der Nederlandse kevers V. Ent. Ber., Amst. 22 (9), 169–171, 1962.

3. Recording unit

To indicate the distribution of the species, we adopted the recording unit of the European Invertebrate Survey (see section 4). We chose the 10×10 km² grid, because a coarser system would have meant the loss of too much information and a finer one would have given excessive scatter due to the lack of sufficient information. Furthermore, the available reports on catch sites generally made the 10×10 km² grid the most suitable, for instance in cases where only the name of the municipality is mentioned.

We found that a species was often recorded more than once in a certain grid. There can be two reasons for this: 1. a collector included two or more specimens caught on a given day at the same site, or 2. the species was caught at different sites in the same grid or on different days at the same site. In the former case we counted the collected individuals as a single record, in the latter as separate records. As an extreme case of the former kind, we considered information from catches performed system- atically at weekly or fortnightly intervals all year round over a varying number of years. Since the total number of catches per species in such programmes is often so large that it would disrupt or bias the distributional pattern of the particular species, all such catches were counted as one record without mention of the dates (see section 7).

On the maps, the number of records per grid is not indicated as in the computer-printed map (see Figure 2) but only the sum total of the records on which the species map is based. As explained in section 7, this latter number is not the same as the one used for the frequency diagram of the seasonal periodicity.

4. Grid system

To locate the data we used the U.T.M. (Universal Transverse Mercator projection) grid system (Anonymous, 1958, Figure 1). The definition of the grids was converted from an alpha-numeric system into a coordinate system to facilitate the processing of the data with a computer (Figure 2). To keep the sizes of the grids constant over the globe, in the U.T.M.-system wedges of variable grid sizes are inserted at regular intervals. Unfortunately, The Netherlands contains one of those wedges, as can be seen in Figure 1. The inserted grids are treated by us as indicated in this Figure.

Other grid systems were available besides the U.T.M. of the European Invertebrate Survey, for instance the national grid currently used by botanists and ornithologists in The Netherlands. For comparative purposes and international studies, however, the disadvantages of a national grid are obvious.

REFERENCE

Anonymous, Technical Manual 5–241–1: Grids and
Grid references. Department of the Army,
Washington 25, Washington DC, U.S.A., 1958.

Fig. 1. The U.T.M. grid system over The Netherlands. Our treatment of the 'wedge' in this system can be understood by comparison of this Figure with the transparent map of the underworked grids (in back pocket). In the latter the whole wedge is delineated, in the former the wedge is truncated at various points. The numbering of the 10×10 km² square within each 100×100 km² square is indicated in the lower left corner.

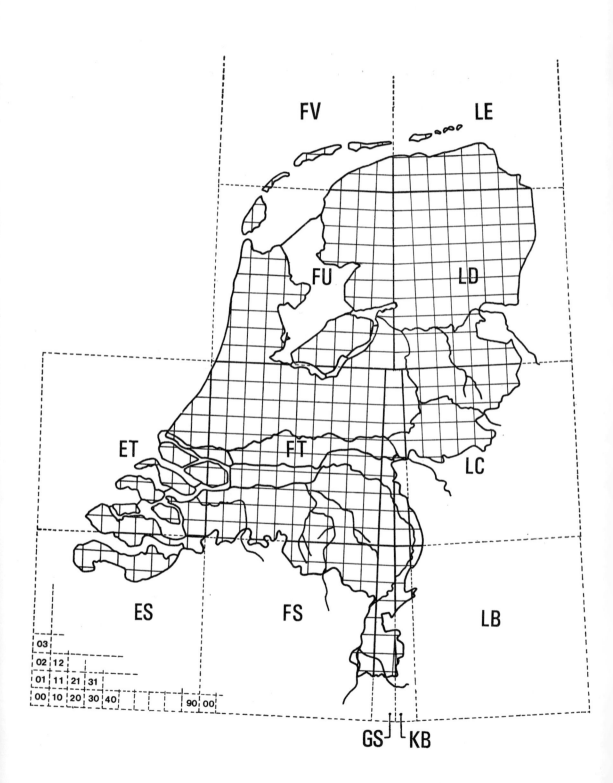

Fig. 2. Computer-printed map for *Elaphrus ullrichi*, illustrating the conversion from the alpha-numeric U.T.M. system to a coordinate system.

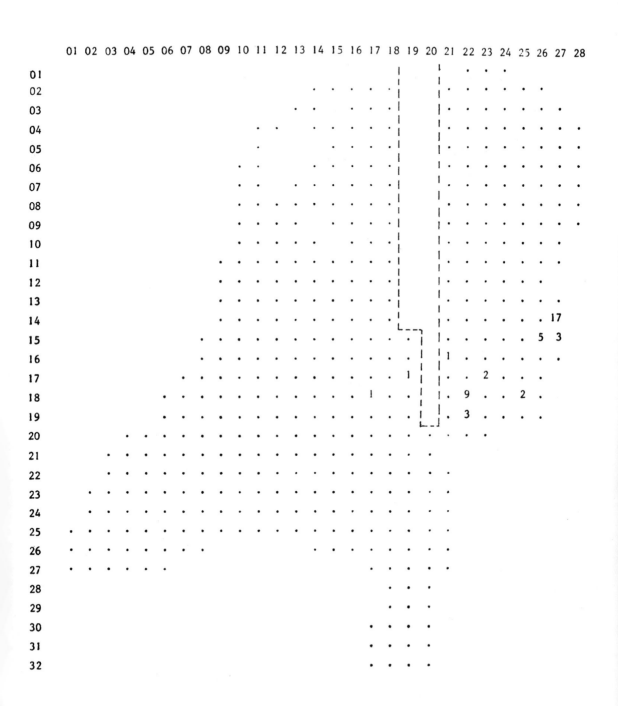

5. Representativity of the material

As a family, carabids are very well known to entomologists, which means that they have been thoroughly collected and identified. Therefore, we assume that the distributional pattern deduced from abundant material may represent the actual pattern, although criteria to evaluate this assumption must of course be developed. If the assumption does not hold, more material must be collected to improve the picture. However, this picture is not necessarily static: a number of species may at a given moment appear among the Dutch beetle fauna, and other species may have disappeared at some time.

We shall first describe the criteria for the representativity of the material in relation to the interpretation of the distributional pattern, and then deal with the dynamic aspects of these patterns with respect to this representativity.

As a measure of representativity we took the number of species recorded per grid, as shown in Figure 3. This Figure gives the bimodal distribution of two groups of data: a first set of roughly 19,000 records (dashed line) and our total set of 51,525 records (solid line).

We reasoned that two factors are involved in the genesis of these bimodal curves: 1. the conditions for a greater or smaller number of species to live in a certain area are not homogeneously distributed over The Netherlands, and 2. the collecting activities are unequally distributed; or both of these two factors combined. The heterogeneous distribution of the living conditions may result in, for instance, a normal frequency distribution of the logarithms of the number of species over the grids. Marked deviations from the latter distribution can result from uneven sampling. Of course, these two effects could be inseparable, in the first place because the heterogeneous distribution of living condition will enhance the unevenness of the collectors' activity.

To evaluate these possibilities, we used the material of our first set of data, originating from the large collection of the Institute of Taxonomical Zoology in Amsterdam together with some smaller collections covering an appreciable part of The Netherlands, to make a map showing the distribution of the numbers of species over the country (Figure 4). Although it is clear from Figure 3 (dashed line) that the two effects in question had certainly played a combined role, it is evident from Figure 4 that too many large parts of the country were underworked. We therefore added data from collections not yet drawn on, paying special attention to local material covering the 'white' parts of the map.

Fig. 3. Number of grids with a certain number of species, according to logarithmic classes. Only the upper limits of these classes are indicated on the abscissa. The arrow indicates the boundary between the underworked grids (on the left) and the well-represented grids (on the right).

Fig. 6. Scattergram showing per species the number of grids in which the given number of records occur.

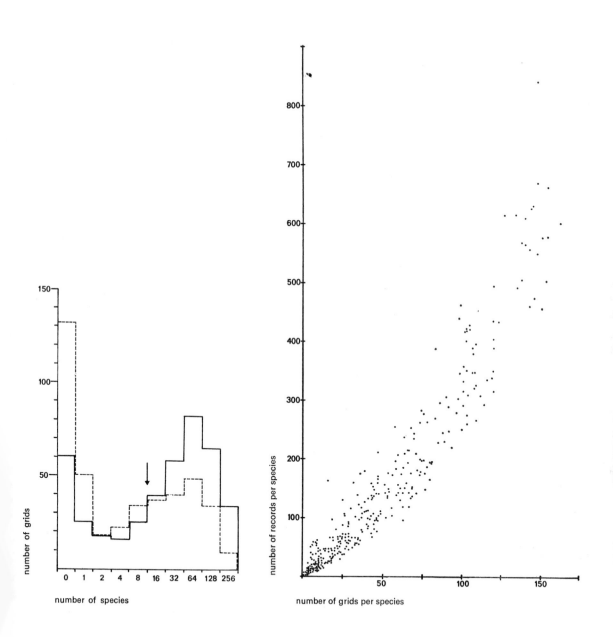

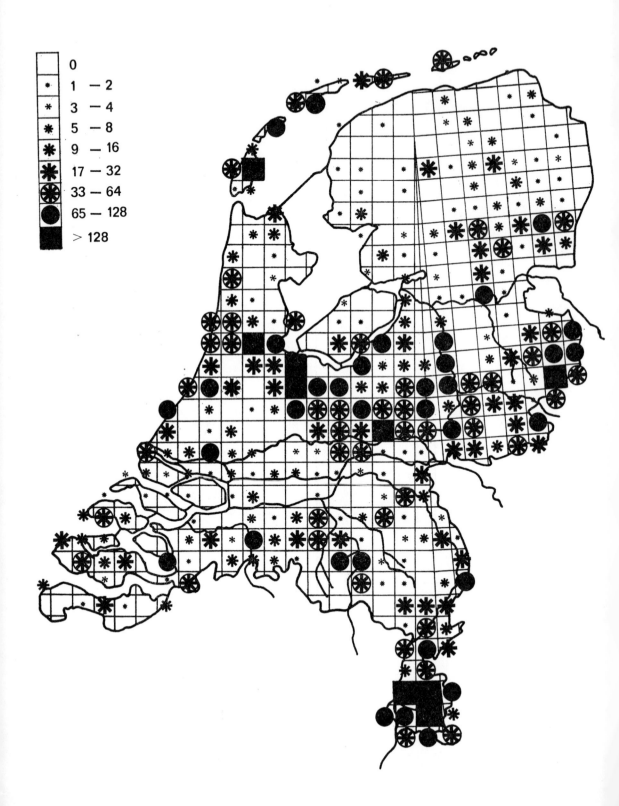

Fig. 4. Number of species per grid, classified in
logarithmic classes on the basis of approximately
19,000 records.

0
1 — 2
3 — 4
5 — 8
9 — 16
17 — 32
33 — 64
65 — 128
> 128

Fig. 5. Number of species per grid, classified in
logarithmic classes on the basis of approximately
51,500 records.

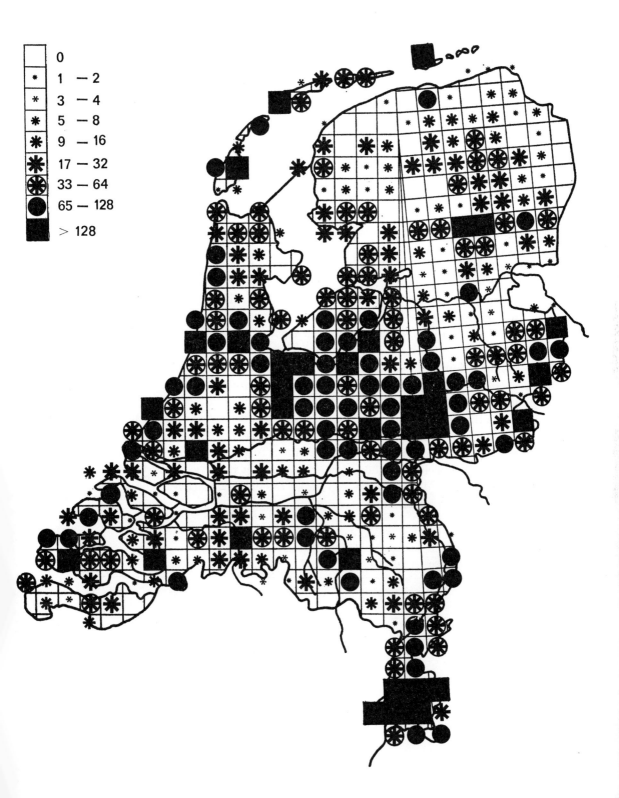

The result is shown in Figure 5, and it gave the solid line in Figure 3. The map shows the more even distribution we expected, and the curve shows a relatively smaller bias toward the lower classes but none in the opposite direction, which means that Figure 5 still has too many underworked grids requiring better sampling by carabidologists. This should be borne in mind in the interpretation of the maps of the individual species; for convenience, we therefore included a transparent map on which the possibly underworked grids are indicated by hatching and which can be laid over the species maps. To determine which grids should be hatched, we divided the bimodal distribution of Figure 3 into two parts: a 'symmetrical' part (classes of 9 to >256 species) to the right of a 'bias' part. We realize that this is an arbitrary division.

We took the number of species as a measure of evenness of the distribution of the collecting density. It could, of course, be argued that the number of records might have been a better measure. If this were true, plotting the number of occupied grids per species against the number of records per species could be expected to give either a curve with one or more points of inflection or a straight curve with a very broad scatter around it, or a combination of the two. In fact, we found a limited scatter around an almost straight line (see Figure 6), which means that the two measures may be used interchangeably.

Since the distributional patterns could be expected to be dynamic, it was necessary to identify not only underworked parts of the country but also changes occurring in the course of time. Table 2, which gives the total number of species per decade, shows a distinct increase in the number of species collected. This table also shows the great variation in the number of records per species.

Table 2. Total number of records and species, as well as the number of records per species, collected in successive decades.

period	decade no.	no. of records	no. of species	no. of records per species
1900	0	1,329	223	5.96
1900–1910	1	2,048	258	7.94
1910–1920	2	4,602	284	16.20
1920–1930	3	4,533	277	16.36
1930–1940	4	3,041	286	10.63
1940–1950	5	2,552	293	8.71
1950–1960	6	4,737	319	14.85
1960–1970	7	11,624	327	35.55
1970–1975	8	11,857	316	37.52
undated		5,202		

Fig. 7. As Fig. 5, but on logarithmic scale, making
it possible to indicate the species-number of the
rare species. We regard as rare 50% of the
species in The Netherlands, i.e. those occurring
in the area in the lower left part of the indicated
50% boundary lines.

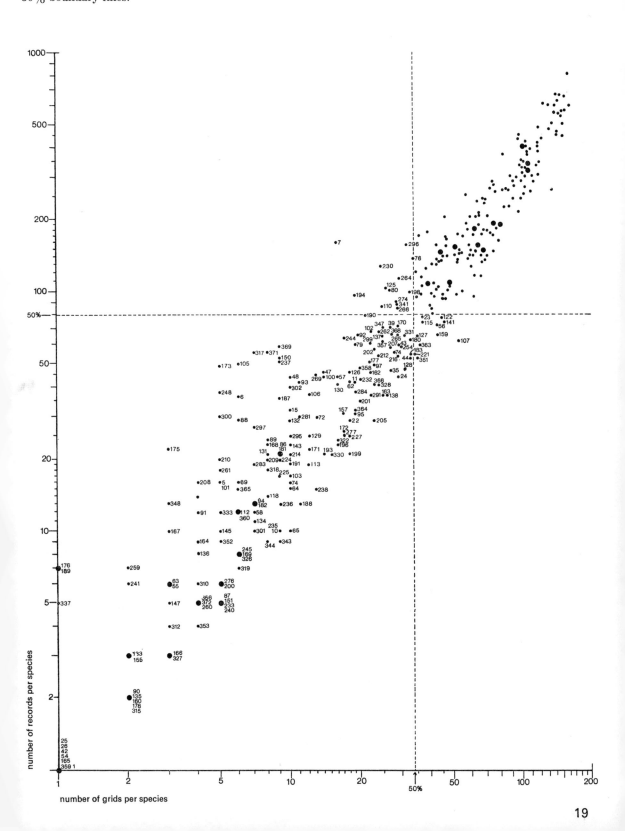

6. Faunistic interpretation of the material: rare and threatened species

With respect to the distinction between rarity and commonness we can ask whether either one actually exists as a biological feature of a species. A biological distinction of this kind should be expressable in a statistical criterion serving to distinguish between the two groups. All discriminating criteria depend on the occurrence or non-occurrence of some kind of discontinuity. However, Figure 6 indicates a continuity with respect to abundance, which means that we cannot separate rare species from common ones on an objective basis. This implies that it is not possible to distinguish patterns in the relative frequencies of rare or abundant species. If, however, we prefer not to abandon these two concepts, we must define them arbitrarily. As a criterion for this we have taken the frequency of 50 per cent below and above a certain value on the basis of the number of records per species in combination with the frequencies of the number of grids occupied per species for a given area during a given period. Figure 7 shows the resulting dividing lines. Thus, according to this criterion, roughly half of the carabid species recorded so far in The Netherlands are defined as rare.

On the basis of this definition, the rarity of a species is dependent on the area covered by collectors as well as on the collecting period. With respect to time we must keep in mind that this definition is static: trends within any given period are not considered. The time factor may, however, be

Table 3. Records of threatened species in the course of time (columns 0–8) and undated records of these together with those of some very rare species (last column).

species	species no.	decade no. (see Table 2)									undated
		0	1	2	3	4	5	6	7	8	
Calosoma sycophanta	24	5	9	2	2	1	–	–	–	–	25
Calosoma auropunctatum	25	–	–	–	–	–	–	–	–	–	1
Calasoma reticulatum	26	–	–	–	–	–	–	–	–	–	1
Notiophilus quadripunctatus	42	–	–	–	–	–	–	–	–	–	1
Dyschirius neresheimeri	54	–	–	–	–	–	–	–	–	–	1
Bembidion striatum	72	2	5	8	–	–	3	–	–	1	14
Tachyta nana	135	–	–	–	–	–	–	–	–	–	2
Chlaenius nitidulus	153	12	3	23	13	9	11	2	4	5	24
Harpalus dimidiatus	187	–	3	3	13	3	5	–	–	1	8
Harpalus modestus	205	5	2	3	2	–	2	–	–	–	15
Dolichus halensis	312	–	–	–	–	–	–	–	–	–	4
Agonum impressum	315	–	–	–	–	–	–	–	–	–	2
Agonum dolens	322	–	2	6	2	–	–	–	–	–	14
Agonum piceum	331	7	5	14	10	2	1	1	8	1	16
Dromius longiceps	348	2	1	1	3	–	–	–	1	–	5

important: for instance, formerly abundant but now rare species are nevertheless defined as abundant. This sometimes means an essential loss of information. Therefore, the definition only applies if there is no trend in the numbers observed during the period under consideration. Species with a negative trend, i.e., for which numbers in a certain area fall below a defined value during the latter part of the total period covered, must be considered threatened. As rare species (both threatened and not threatened) we consider those indicated by their number in Figure 7. Threatened species are listed in Table 3. For some very rare species no collecting date was available; the number of records of these species are also enumerated in Table 3 (last column).

7. Seasonal periodicity

Collecting dates obtained from the labels attached to specimens in collections were punched on cards, which could then be sorted according, for instance, to the month of capture. From these dates a frequency diagram was constructed for each species. This diagram is included on the map. Because the collecting date does not occur on some labels and is rarely mentioned in the literature, the total frequency in this diagram does not agree with the number of records mentioned in section 3. Furthermore, since the collecting dates of material obtained in research programs in some places in The Netherlands are also unknown to us (see section 3), these catches too were excluded from the determinations of the seasonal periodicity.

The catching of a beetle depends on three factors: its presence at a given place on a given day, the activity of the collector, and the sampling technique used. The degree to which collectors are active in certain parts of the year may influence the picture of the seasonal periodicity of a species (see Den Boer, 1967). Although the systematic catches made by means of pitfall traps in research projects on the ecology of carabid beetles give more adequate information than the data from museum records and from the literature, we used only the latter. We did so because these data originated from material covering a long period and were collected in all parts of The Netherlands. In this way, we arrived at a picture based on one type of collecting only.

The diagrams based on these data are restricted to adults, because only this information was extracted from the collections. To make the information on the seasonal periodicity of the larvae given by Larsson (1939) more accessible, it is given here as well, but because it is based on Danish material, which may differ from the Dutch, it is not shown quantitatively as a frequency diagram but qualitatively by bars. It remains possible that even this qualitative picture is biased in this or some other way.

REFERENCES

Boer, P. J. den, De relativiteit van zeldzaamheid. Ent. Ber., Amst. 27, 52–60, 1967.

Larsson, S. G., Entwicklungstypen und Entwicklungszeiten der dänischen Carabiden. Ent. Medd. 20. København, 1939.

8. European distribution

Because the distribution of the carabid beetles over The Netherlands almost always forms only a very small part of the distribution for Europe as a whole, we thought it might be of interest to include the latter. These maps, which are based on the literature cited below, make it possible to classify a species as, for instance, a northern one or a purely coastal one, and show whether its occurrence in The Netherlands is near the outer limit of its range of distribution.

The survey of the literature gave us the impression that the amount of information provided by the records underlying the European distribution pattern decreases from west to east and from north to south. For practical reasons, we restricted this part of the study to the literature on the distribution as far eastward as the eastern border of European Russia. Since North Africa is considered to belong to the same zoogeographic region as Europe, we included data on the distribution north of the Sahara as well.

REFERENCES

Apfelbeck, V., Die Käferfauna der Balkanhalbinsel 1. Berlin, 1904.

Bonadona, P., Catalogue des Coléoptères carabiques de France. Toulouse, 1971.

Burakowski, B., M. Mroczkowski, J. Stevanska, Katalog Fauna Polski, 13,2, Coleoptera-Carabidae. Warszawa, 1973.

Burmeister, F., Biologie, Ökologie und Verbreitung der europäischen Käfer 1. Krefeld, 1939.

Csiki, E., Die Käferfauna des Karpaten-Beckens 1. Budapest, 1946.

Derenne, E., Catalogue des Coléoptères de Belgique 2, Caraboidea, Carabidae. Bruxelles, 1957.

Everts, E., Coleoptera Neerlandica 1, 3. Den Haag, 1898, 1922.

Fuente, J. M. de la, Catalogo sistematico-geografico de los Coleopteros en la peninsula iberica etc. Bol. Soc. Ent. España, 1–3, Zaragoza, 1918–1921.

Horion, A., Faunistik der deutschen Käfer, 1, Adephaga-Caraboidea. Wien, 1941.

Jacobson, G., The beetles of Russia and Western Europe (in Russian), Leningrad, 1905, 1908.

Lindroth, C. H., Die Insektenfauna Islands und ihre Probleme. Zool. Bidr. 13, 103–599, Uppsala, 1931.

Lindroth, C. H., Die Fennoskandischen Carabidae 1, 3. Göteborgs K. Vetensk. o. Vitterh. Samh. Handl. (6) B, 4, Göteborg, 1945, 1949.

Lindroth, C. H., Handbooks for the identification of British insects. Coleoptera, Carabidae. Roy. Ent. Soc. London, IV, 2, 1974.

Magistretti, M., Coleoptera: Cicindelidae, Carabidae – Catalogo topografico. Fauna d'Italia, 8. Bologna, 1965.

Moore, B. P., The British Carabidae (Coleoptera), Part II: The county distribution of the species. Entomologist's Gaz. 8: 129–137, 1957.

Netolitzky, F., Die Verbreitung von Bembidion tibiale. Ent. Blätter 8 (1), 1912.

Netolitzky, F., Die Verbreitung von Bembidion atrocoeruleum. Ent. Blätter 8 (4/5), 1912.

Netolitzky, F., Die Verbreitung von Bembidion striatum. Ent. Blätter 14 (4/6), 1918.

Netolitzky, F. and P. Meyer, Die Verbreitung von Bembidion illigeri. Ent. Blätter 23 (3), 1927.

Netolitzky, F. and P. Meyer, Die Verbreitung von Bembidion argenteolum. Ent. Blätter 29, 1933.

Netolitzky, F. and P. Meyer, Die Verbreitung von Bembidion fluviatile. Ent. Blätter 35 (1), 1939.

Rathlef, H. von, Coleoptera Baltica. Käferverzeichnis der Ostseeprovinzen. Arch. f. Naturk. Liv-, Est- und Kurlands (II. Ser.) 12. Dorpat, 1905.

Sainte-Claire Deville, J., Quelques aspects du peuplement des Iles britanniques (Coléoptères). Soc. Biogéogr. 3, Paris, 1930.

9. Wing form

Because in our opinion information concerning the wings of carabids (i.e., whether all individuals of a species are macropterous or brachypterous) is very useful, it is given on the species maps. Most of this information is taken from Lindroth (1945, 1974). Where it concerns deviations from the Scandinavian situation and species not mentioned by Lindroth, we used information from Den Boer pers. comm. and from our own observations. In the cases in which information was completely lacking we omitted the wing indication instead of updating the Dutch material on this point.

REFERENCES

Lindroth, C. H., Die fennoskandischen Carabidae I. Göteborgs K. Vetensk. o. Vitterh. Samh. Handl. (6) B, 4, Göteborg, 1945.

Lindroth, C. H., Handbooks for the Identification of British Insects. Coleoptera, Carabidae. Roy. Ent. Soc. London, IV, 2, 1974.

10. Sources

1. COLLECTIONS

a. *Museum collections*

Amsterdam, Institute of Taxonomical Zoology; comprising the collections of: K. J. W. Bernet Kempers, J. Broerse, A. J. Buis, J. B. Corporaal, F. C. Drescher, D. van der Hoop, H. J. Klaassen, D. MacGillavry, A. C. Nonnekens, A. van de Post, P. van der Wiel and C. Willemse.

Enschede, Natural History Museum.

Leiden, National Museum of Natural History; comprising the collections of: P. J. Brakman, E. Everts, C. J. Ruurs, and F. T. Valck Lucassen.

Wageningen, Department of Entomology, Agricultural University; Uyttenboogaart collection.

Leeuwarden, Frisian Natural History Museum; collection of G. Stobbe.

b. *Private collections*

Blokland, A. (card index), Heerlen
Boer, P. J. den (card index), Wijster
Booy, K. (card index), Wageningen
Etten, J. van, Amsterdam
Gijzen, T. van, Arnhem
Heijnsbergen, S. van, Naarden
Heyerman, Th., Randwijk
Jongema, Y., Wijster
Littel, A., Amsterdam
Stuivenberg, F. van, Enschede
Teunissen, A., Eindhoven
Turin, H., Oosterbeek
Vester, K., Emmen
Veldkamp, W. J., Lichtenvoorde

2. SYSTEMATIC SAMPLING

a. *Biological Station, Wijster* (P. J. den Boer)

Benderseweg (woods), 1959–1961
Biological Station, 1966–1967
Bruntinge (woods), 1962–1966
Dalerpeel (peat marsh), 1962–1963
Dwingelo (heath), 1959–1960
Dwingelo (plantations), 1959
Eursinge (woods), 1960
Hullenzand (inland dunes), 1963–1967
Kampsweg (peat marsh), 1959–1961
Kampsweg (woods), 1959
Kibbelhoek (woods), 1962
Kibbelhoek (plantation), 1960–1966
Kralo (arable land), 1965
Kralo (coppice), 1962–1966
Kralo (heath), 1959–1967
Lheebroek (inland dunes), 1959
Mantinge (inland dunes), 1965
Mantinge (woods), 1959–1966
Meerstalblok (peat marsh), 1961
Noordlagen (woods), 1962–1965
Onland Lheebroek (wet meadows), 1960
Poort II (peat marsh), 1960
Terhorsterzand (inland dunes), 1959)1963
Turfveen (peat marsh), 1963

b. *Institute for Ecological Research, Arnhem*

Noordoost Polder: two woods 1968, road verges 1972, 1973
Oostelijk Flevoland: reed stands, arable fields, plantations, and road verges 1968–1975
Zuidelijk Flevoland: various places in the reed swamps, arable fields, plantations, and road verges, 1968–1975
Markerwaard, dike, 1968, 1975
Islands in Veluwemeer, 1968, 1969
Islands in Grevelingen, 1973–1975
Border region of the IJsselmeer, road verges, 1973
Friesland, Noord-Holland, Zuid-Holland, road verges, 1974
Afsluitdijk, verges, 1974
Nijkerk, wood and meadow, 1968–1970

c. *Institute of Phytopathological Research, Wageningen* (W. C. Nijveldt)

Oostelijk and Zuidelijk Flevoland: arable fields, 1974, 1975

d. *Research Institute for Nature Management* (R.I.N.), *Arnhem* (J. van der Drift)

Grubbenvorst, 1953, 1958, 1960
Hackfort, 1957–1960

Hoge Veluwe, 1951–1953
Noordoost Polder, 1965

*e. IJsselmeer Polders Development Authority,
Lelystad* (M. Zijlstra)

Grevelingen, 1972–1975
Oosterschelde, 1974–1975

f. Zoology Department, Free University, Amsterdam
(J. Meijer)

Lauwerszee and surroundings, 1969–1975

g. Zoology Department, University of Leiden
(G. J. de Bruijn)

Meyendel: dune area, 31 sites, 1953–1960

---

3. DATA FROM THE LITERATURE

*3.1. Card index of P. van der Wiel for the Frisian
Islands*

(Tx=Texel; V=Vlieland; T=Terschelling;
A=Ameland; S=Schiermonnikoog; R=Rottum;
G=Griend.
TVE=Tijdschr. Ent.; EB=Ent. Ber., Amst.)

a. Published literature

Everts, E., 7e Wintervergadering. TVE 17, 1874.
Everts, E., Suppl. lijst Ned. Col. TVE 24, 1881
Everts, E., Nieuwe Naamlijst Ned. Schildvl. Ins., 1887
Everts, E., 5e lijst van soorten etc. TVE 51, 1908
Everts, E., 8e lijst van soorten etc. TVE 55, 1912

Kempers, K. J. W., Bijdrage Col. Fauna van Texel. TVE 40, 1897

MacGillavry, D., Entom. Fauna eil. Terschelling. TVE 57, 1914

Reclaire, A., Meded. Kevers op Terschelling. EB 7, 1926
Reclaire, A., Meded. Kevers op Vlieland. EB 8, 1930
Reclaire, A., Twee korte meded. Vlieland. EB 8, 1932
Ritsema, C., Ann. Soc. Ent. Belg. 15, 1872
Ritsema, C., 27e Zomerbijeenk. TVE 16. (Tx, V, T), 1873
Ritzema, J., Bijdr. Entomofauna Noordzeeëil. TVE 16, 1873

Veth, H. J., 12e Winterverg. TVE 22. (T), 1879

*b. Personal communications received by
P. van der Wiel*

Everts, E.: List of sampling localities for 1925–1932.
Booy, H. L., 1937. T
Broerse, J., 1924. Tx, A
Burger, F. W., 1917. Tx
Companjen, A., 1936. G
Doeksen, J., 1937. T
Evers, A. M. J., 1936. T

Gravestein, W. H., 1935–37, 39, 46, 47. Tx
Hoop, D. van der, 1901. S
Kabos, W. J., 1938. Tx
Kempers, K. J. W., 1896. Tx
Klynstra, D. H., 1939. Tx, T
Kruseman, G., 1938. G
Lieftinck, M., 1921. Tx
MacGillavry, D., 1912. T
MacGillavry, H. J., 1925. T
Nieuwenkamp, M., 1928. A
Nonnekens, A. C., 1922–24. Tx
Nonnekens, A. C., 1937. T
Ooststroom, S. J. van, 1935. A, Tx
Reclaire, A., 1929–31. V
Reclaire, A., 1925, T
Reclaire, A., 1946–48. Tx
Ritsema, C., 1872. Tx, V
Ritsema Bos, J., 1871. Tx, T, A, S, R
Uyttenboogaart, D. L., 1939. Tx
Valck Lucassen, E. T., 1926. Tx, V, T
Valentin, A. C. van, 1935. Tx, T
Veth, H. J., 1878. T
Wiel, P. van der, 1921, 33, 39, 46–48. T
Wiel, P. van der, 1931. V

*3.2. Inventory reports in the files of the Research
Institute for Nature Management (R.I.N.), Leersum*

Amstelveen, Amsterdamse Bos, Med. Centr. Inst. v. Flor. Ond. 8, 195, 1948.
Bemelen, Bemelerberg, 1966: P. Poot
Boxtel, De Geelders, 1966: P. Aukema
Colmont, Wrakelberg, 1960–1965: P. Poot
De Wijk, De Reest, 1966: C. D. Kruizinga
Echt, De Doort, 1968–1970
Enschede, Aamsveen, 1964: N.J.N. afd. Enschede
Ermelo, Speulder en Sprielder bossen, 1962: W. Reijnders
Ginneken, De Goudberg, 1967: P. Aukema
Gorssel, Gorsselse Heide, 1969: C.J.N., contr. C. van Houdt

Gronsveld, 1966: P. Poot
Meyel, De Grote Peel, 1965
Millingen, Millinger Waard, 1956: C. J. Hogendijk
Naarden, 1950: R. Tolman
Neede, Needse Achterveld, 1965: P. Poot
Ophemert, (woods) 1956: C. J. Hogendijk
Rossum, Kil van Hurwenen, 1956: C. J. Hogendijk
Soest, 1949, 1950: R. Tolman
Tegelen, Holtmühle, 1967: N. P. Peerdeman
Terschelling, Texel, Vlieland, 1937, 1938: Inventory
State-owned nature reserves
Ubbergen, Duivelsberg, 1964: E. Mols
Valkenburg, Ravensbos, 1966: P. Poot
Vianen, Bos van Vianen, 1956: C. J. Hogendijk
Wijhe, Duursewaarden, 1956: C. J. Hogendijk
Wijlre, (woods), 1966: P. Poot

3.3. Carabid inventories by youth societies for the study of nature, mostly unpublished

Drentse A, 1969, 1970, 1971: c.j.n., Anax
Gorsselse Heide, 1970: c.j.n., Anax 71
Ratumse Beek, Winterswijk, 1971
Schiermonnikoog, 1956: n.j.n.
Schiermonnikoog, 1969, 1970: c.j.n.
Terschelling, 1971, 1972: c.j.n. + n.j.n.
Vlieland, 1971, 1972: n.j.n.
Zutphen, wet meadows

3.4. Additional literature

Batten, R., Enkele notities over het voedsel van Cicindela en een nieuwe vindplaats. Ent. Ber., Amst. 14, 1953.

Berger, C. G. M. and P. Poot, Nieuwe en zeldzame soorten voor de Nederlandse keverfauna I. Ent. Ber., Amst. 30, 213–214, 1970.
Berger, C. G. M. and P. Poot, idem II. Ent. Ber., Amst. 32, 26, 1972.

Boer, P. J. den, Activiteitsperioden van de loopkevers in Meyendel. Ent. Ber., Amst., 8, 80–88, 1958.

Booy, K., Inventory of the outer marches of the Rhine near Wageningen, 1969–1972, 1972.

Brakman, P. J., Bradycellus sharpi, een nieuwe carabide voor de Nederlandse fauna. Ent. Ber., Amst. 11 (260), 282, 1944.
Brakman, P. J., Korte Coleopterologische notities. Ent. Ber., Amst. 13, 250, 1951.
Brakman, P. J., Ent. Ber., Amst. 15, 181–182, 1954.

Bruyn, H. de, Loopkevers in opgespoten land. Amoeba 43, 111–120, 1967.

Doesburg, P. H. van, Carabidae and Staphylinidae from Terschelling 1950–1952 and Grubbenvorst 1953–1955. Internal report Zool. Dep. Univ. Utrecht, 1958.

Eyndhoven, G. L. van, Verslag van de 116e zomerbijeenkomst van de Ned. Ent. Ver. Ent. Ber., Amst. 22 (7), 1962.

Gijzen, T. van, Coleoptera inventory Linschoterbos. Internal report Zool. Dep. Univ. Utrecht, 1970.

Heerdt, P. F. van, J. Isings and L. E. Nijenhuis, Temperature and humidity preferences of various Coleoptera from the duneland area of Terschelling. Proc. Kon. Ned. Akad. v. Wetensch. part 1, C 59 (5), 668–676; part 2, C 60 (1), 99–106, 1956–57.

Hengst, J., Loopkevers in de duinen. Amoeba 32 (5), 1956.

Heijnsbergen, S. van, Carabids from Het Noordhollands Duinreservaat from 1939–1955.

Klynstra, B. H., Mededelingen over de Nederlandse Adephaga (Col.) III, Ent. Ber., Amst. 318, 369–372, 1951.
Laboratory of Applied Entomology Amsterdam, 1963. Excursion to Voorne.

Mooy, W. A. M., Zoölogisch oecologische studie van de lopende keverfauna in het oostelijk deel van het Krommerijngebied. Kromme Rijn Rapport 7, 1973.

Nonnekens, A. C., Iets omtrent de keverfauna van het Amsterdamse Bos. Ent. Ber., Amst. 16 (7), 1956.

Reimerink, H. G. A., Kolonisatie door Carabidae in het gebied van De Knar. Internal report Institute for Ecological Research, Arnhem, 1970.

Terpstra, B., Keveronderzoek in het Krommerijngebied. Kromme Rijn Rapport 15, 1974.

Twentse Insekten Werkgroep, Carabids in Twickel, 1972.

Walrecht, B. J. J. R., Carabus coriaceus in Zeeland. De Levende Natuur 239, 1963.

Wiel, P. van der, Bijdrage tot de kennis der Nederlandse kevers, V, Ent. Ber., Amst. 22, 169–171, 1962.

3.5. Personal communications

a. Sites sampled with pitfall traps

site	year	collector	1	2	3	4	5	6	7	8
Aalten, 5 sites	1973	T. Heyerman			3					
Achlum, various sites	1974/75	A. de Smidt	1			4				
Arnhem, southeast	1974/75	R. Dorré					5			
Arnhem, southwest	1974/75	T. van Gijzen				4				
Amsterdam, Bijlmer	1973/74	A. Littel							7	
Braakmanpolder	1972	H. Kooman	1							
Doorwerth, wood	1971	H. Turin	1					6		
Dwarsgracht	1974	J. Nip	1			4				
Ede, Planken Wambuis	1973	K. Alders	1		3					
Enschede, Hof Espelo	1970/71	H. Turin						6		
Enschede, Lonnekerberg	1970/71	H. Turin						6		
Follega,	1973/74	R. van de Ree	1			4				
Gorssel, heath	1972/73	W. Verholt	1	2	3			6		
Hemmen, wood	1974	T. Heyerman			3					
Hoge Veluwe	1972/74	T. Heyerman	1		3					
Den Haag, dunes	1973	G. Speek	1							
Kennemerduinen, many sites	1973/74	K. Alders	1							
Naardermeer, 5 sites	1974	W. van Wijngaarden	1							
Oostvoorne, dunes	1972/73	T. Heyerman+K. Alders	1		3					
Pannerden, outer marches	1972	K. Booy								8
Randwijk, garden	1974	T. Heyerman			3					
Schiermonnikoog	1973/74	A. Littel							7	
De Steeg, outer marches	1973	T. Heyerman	1		3					
De Steeg, hedges	1972/73	T. Heyerman	1		3					
De Steeg, Faisantenbos	1974	T. Heyerman	1		3					
Stroe, many sites	1974/75	T. van Gijzen	1			4	5			
Terlet	1973/74	K. Alders	1							
Tynaarlo en Kogelbergerveen	1975	A. Rijnsdorp			3					
Westeinderplassen	1971	K. Bol						6		
Wijhe, Duurserwaarden	1971/73	W. Gerritse	1	2	3					
Winterswijk, Bekkendelle	1973	T. Heyerman			3					
Winterswijk, Vlist	1973	T. Heyerman			3					
Zutphen, outer marches	1970/72	K. Booy								8
Deventer, Rande, forest	1971/73	W. Gerritse	1	2	3					

b. Sites sampled by hand

site	year	collector	site	year	collector
Aalten, arable land	1973	3	Nijmegen, ruderal site	1971	1, 14
Aalten, Heurne	1973	3	Ooypolder	1973	3
Arnhem, Inst. Ecol. Res.	1974	3, 5	Oostvoorne	1971	1
Bedum (Gr.)	1972/73	8	Oostvoorne, dunes	1972	1, 3, 6
Drentse A	1969/70	9	Ossendrecht	1973	1
Dijkwater Schouwen	1973	3	Otterlo, heath	1972	1, 3
Ede, Planken Wambuis	1972/73	1, 3	Schaarsbergen, wood	1971	1
Ermelo	1973	1, 3	Schaarsbergen, ruderal	1971/72	1, 6
Garderen, arable land	1974	1, 3	Schiermonnikoog	1971/72	13
Goeree, Punt	1973	1, 3	Staverden	1974	1, 3
Grevelingen, Hompelvoet	1973	1, 3	De Steeg, Havikerwaard	1973	3
Haarlem, many sites	1973/75	1	Terlet	1971/73	1, 3
Hellendoorn	1973	1, 3	Terschelling, Bosplaat	1972	1, 14
Heveadorp	1972	3	Texel	1971/72	11
Heteren, arable + dike	1973/74	1, 3	Wageningseberg	1972	1, 3
Hoge Veluwe	1972	1, 3	Weerribben	1972	12
Huissen	1971	3	Winterswijk, Bekkendelle	1973	3
Kennemerduinen	1973	1	IJsselmeerpolders	1971/72	1, 2, 10, 14
Linschoten	1969/70	4	Zeeuws Vlaanderen	1973	3
Nijkerk, Meerte	1973	3	Zuid-Limburg, many sites	1972	1, 3, 6

The numbers indicate:

1 = K. Alders	5 = R. Dorré	9 = S. G. Dutmer	13 = J. A. Kooman
2 = H. Krols	6 = H. Turin	10 = H. Reimerink	14 = K. Reinink
3 = T. Heyerman	7 = A. Littel	11 = G. van Noort	
4 = T. van Gijzen	8 = K. Booy	12 = F. v. Stuivenberg	

Acknowledgements

We are greatly indebted to all those who contributed to this work by supplying us with data from their collections or by admitting us to the museum collections. In addition, we wish to thank a number of people for special assistance. A. Turin-van den Burg gave invaluable help with coding and punching and also drew parts of the maps. J. A. van Noort-Kooman, K. Booy, H. Krols, K. Reinink, and F. van Stuivenberg spent much time in collecting data from museums and private collections. P. J. den Boer, J. van der Drift, and G. J. de Bruyn supplied us with data from their extensive material collected by systematic sampling. K. Alders, R. Dorré, T. van Gijzen, and Th. Heyerman identified large numbers of specimens collected in our own systematic sampling in the IJsselmeer-polders and at a number of sites in underworked areas. L. D. B. van den Burg designed several sorting programs. T. Tjebbes added the grid characteristics to punch cards. B. J. Turin made a survey of the first 19,000 records and prepared a preliminary series of maps based on that part of the material. L. Lobach punched innumerable cards in a very short time without losing his enthusiasm for the work. J. H. H. Lochtenberg prepared most of the maps. Mrs. I. Seeger read the English text. F. C. Bos designed the dustjacket.

Explanation of the maps

Species names: according to Brakman (1966)

Species names between parentheses: according to Lindroth (1974)

Species numbers: according to Brakman (1966) (except number 359.1, indicating an additional species of the genus *Dromius*)

Wing form: according to Lindroth (1945), Lindroth (1974), Den Boer (pers. comm.), and our own observations

Bar for larvae: seasonal presence, according to Larsson (1939)

Number of records: total number of records on which the map is based (usually greater than the number of dots on the maps)

Frequencies in the histograms: number of specimens per month, as found in museum collections, private collections, or the literature (undated specimens and specimens from ecological investigations were not included)

Maps showing the European distribution in black: based on the literature

Symbols:

↘ indicates occurrence on very small islands or isolated sites outside the main area
? doubtful occurrence
+ extinct

Transparent maps

1. Underworked grids
Hatching: grids with less than 9 species (see page 15, Fig. 3).

2. Phytogeographical regions of The Netherlands according to J. L. van Soest, adopted from: Atlas van Nederland, plate vi–3, Den Haag, Staatsuitgeverij, 1976.

3. Land use in The Netherlands
adapted from: Landbouwatlas van Nederland, Zwolle, Tjeenk Willink, 1959.

4. Soil map of The Netherlands
in very simplified form adapted from: Atlas van Nederland, plate iv–12, Den Haag, Staatsuitgeverij, 1964.

Maps

1 Cicindela silvatica L. *C. sylvatica L.*   macropterous 181 records

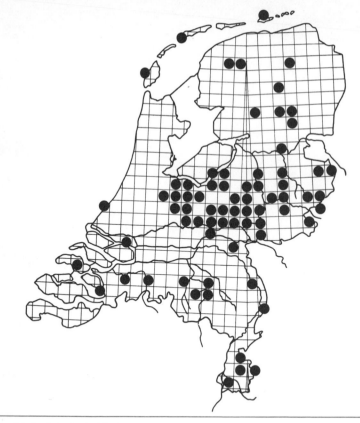

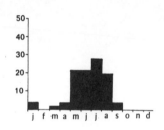

2 Cicindela hybrida L.   macropterous 350 records

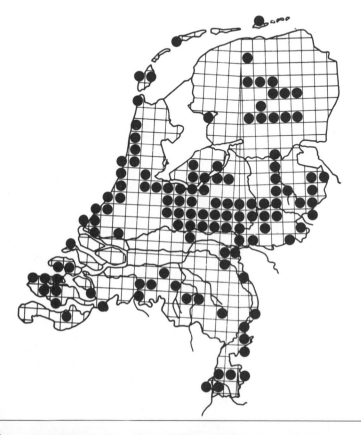

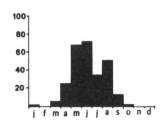

3 Cicindela maritima Dej.

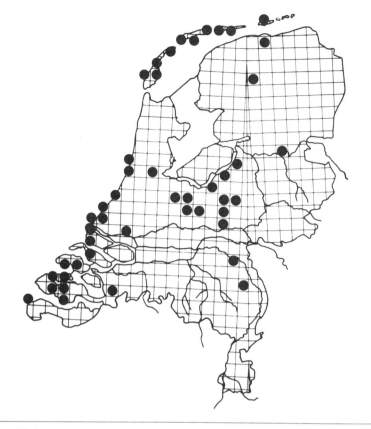

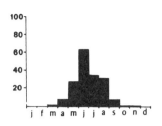

4 Cicindela campestris L.

macropterous   396 records

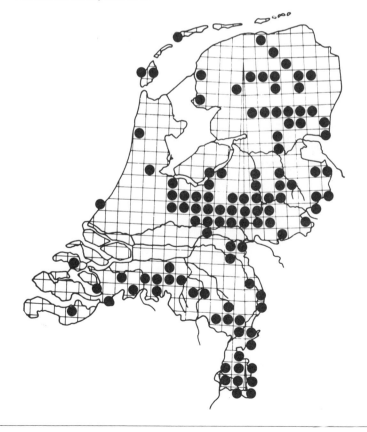

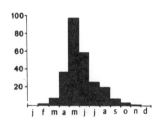

5 Cicindela germanica L.

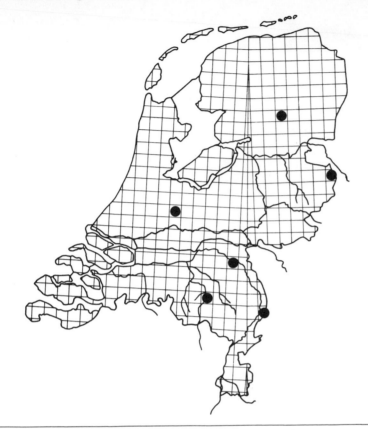

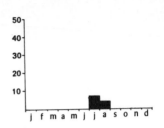

6 Cicindela trisignata Dej.

macropterous 37 records

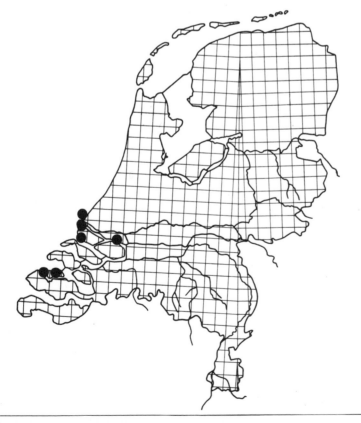

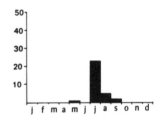

7 Cychrus caraboides (L.)

brachypterous 162 records

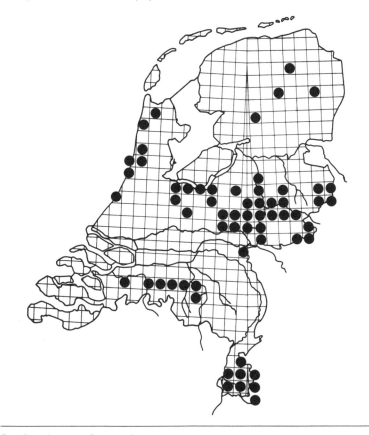

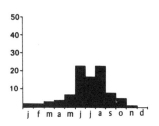

larvae

8 Carabus coriaceus L.

brachypterous 115 records

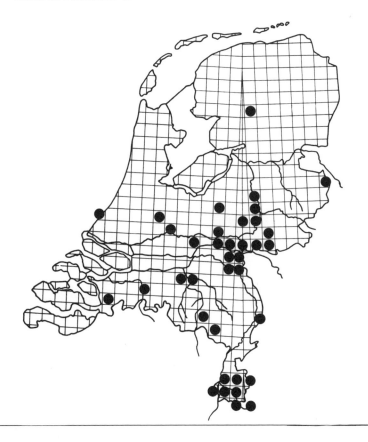

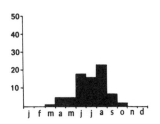

larvae

9 Carabus purpurascens F.

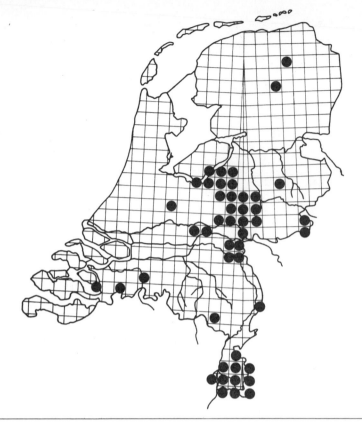
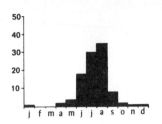

10 Carabus intricatus L.

brachypterous   10 records

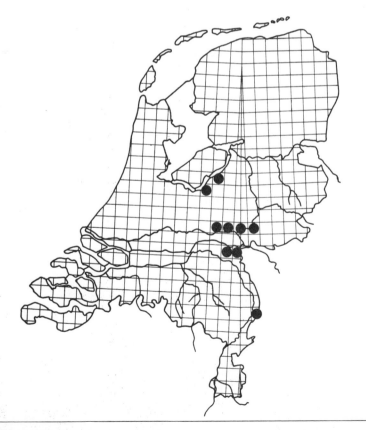
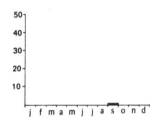

larvae

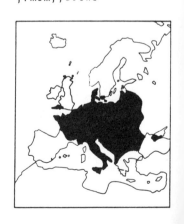

11 Carabus auronitens F.

brachypterous 42 records

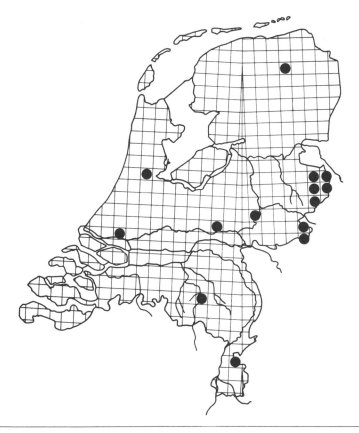
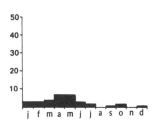

12 Carabus problematicus Hbst. *C. catenulatus auct.*

brachypterous 388 records

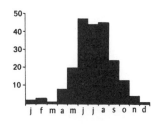

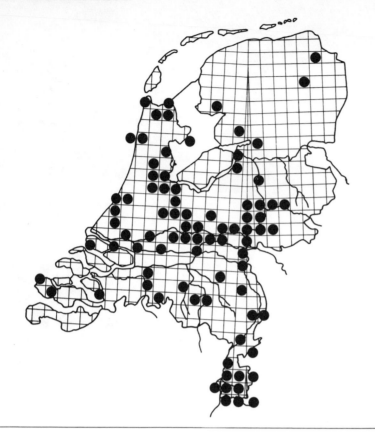

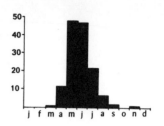

14   Carabus granulatus L.   dimorphic   434 records

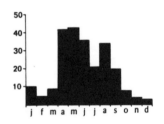

larvae

15 Carabus convexus F.

brachypterous   32 records

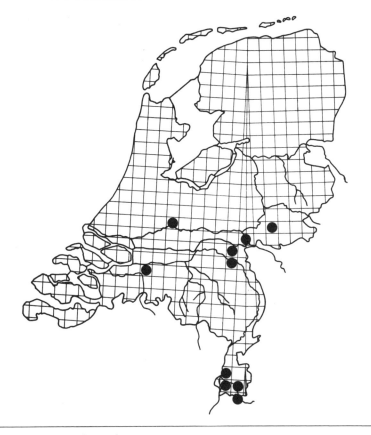

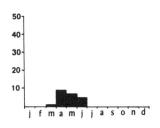

16 Carabus nitens L.

brachypterous   209 records

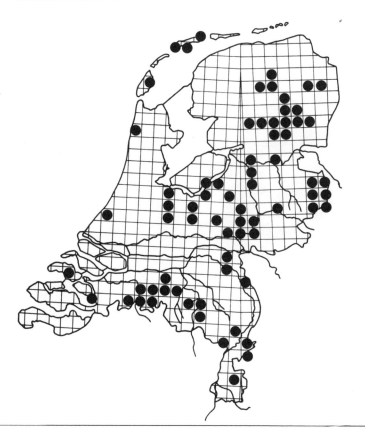

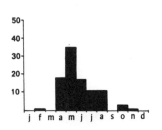

larvae

17 Carabus clathratus L.

dimorphic 110 records

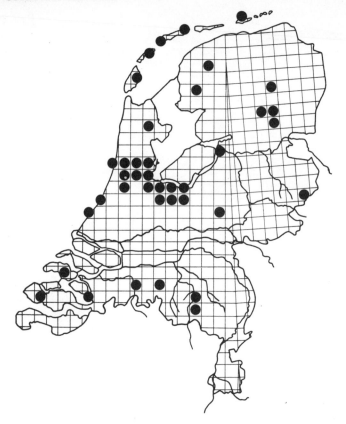

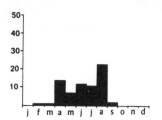

larvae

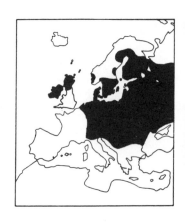

18 Carabus cancellatus III.

brachypterous 234 records

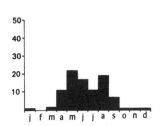

larvae

19 Carabus arcensis Hbst. *C. arvensis Hbst.*   brachypterous   239 records

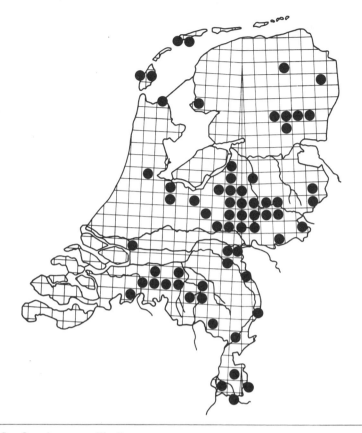
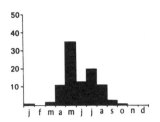

20 Carabus monilis F.   brachypterous   131 records

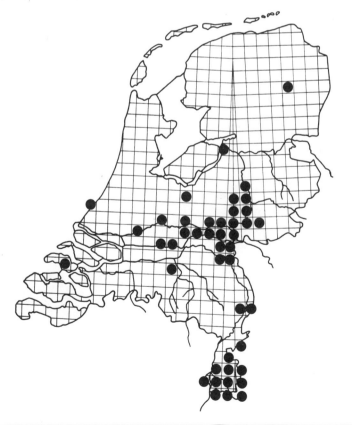
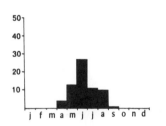

21   Carabus nemoralis Müll.

brachypterous   417 records

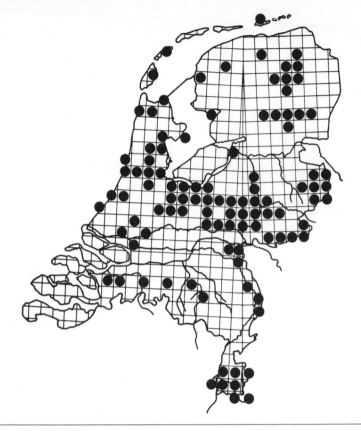

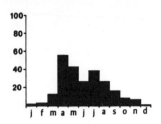

22   Carabus glabratus Payk.

brachypterous   29 records

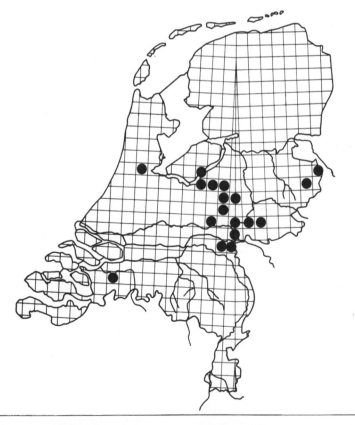

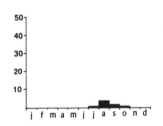

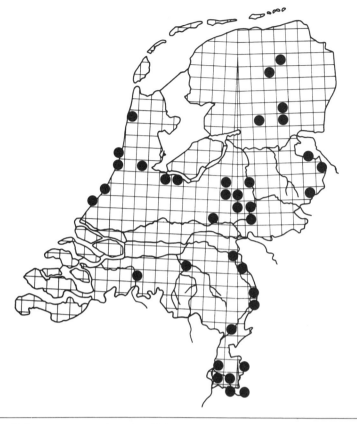

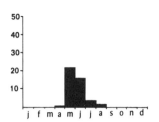

larvae

24 Calosoma sycophantha (L.)   macropterous   44 records

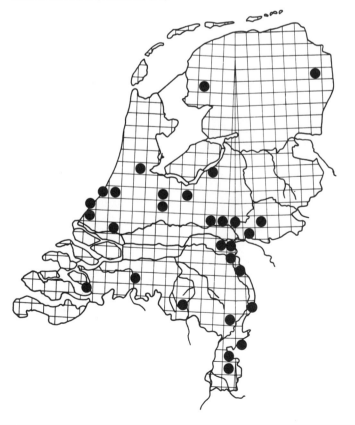

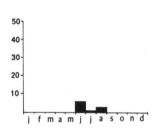

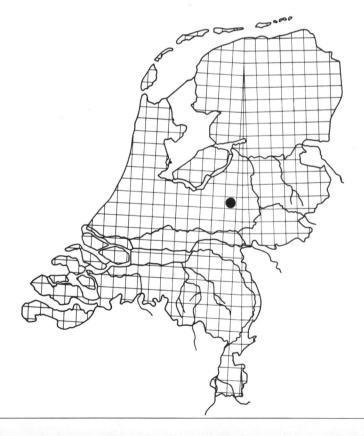

27 Leistus spinibarbis (F.)

macropterous   133 records

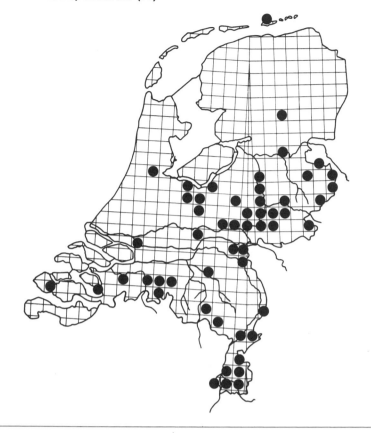

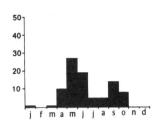

28 Leistus rufomarginatus (Dft.)

macropterous   206 records

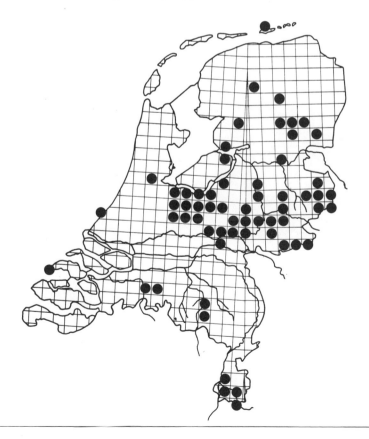

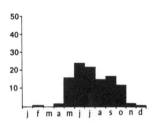

larvae

29 Leistus fulvibarbis Dej.

71 records

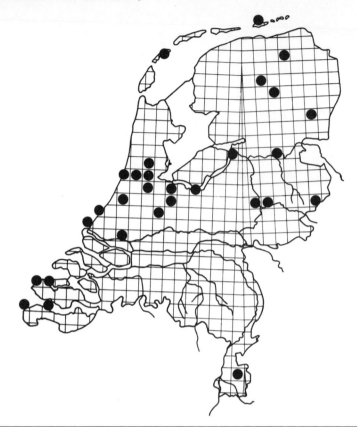

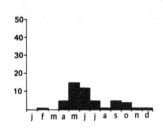

30 Leistus rufescens (F.)

dimorphic   359 records

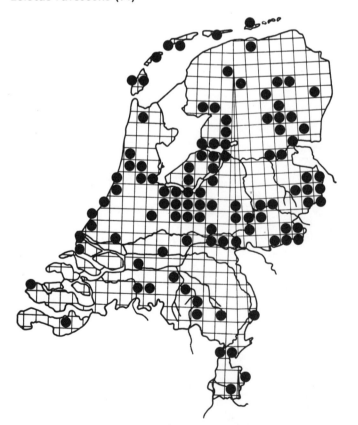

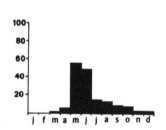

larvae

brachypterous   184 records

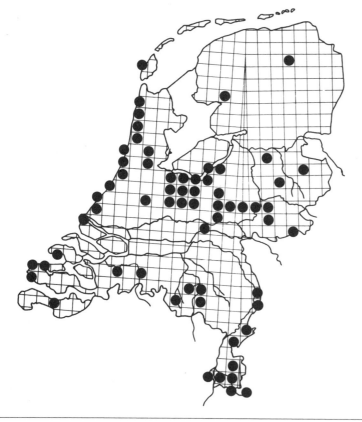

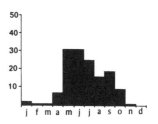

macropterous   95 records

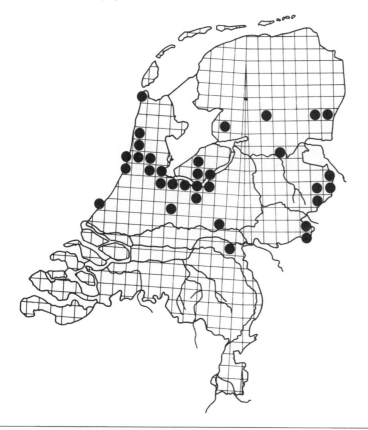

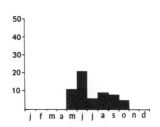

33 Nebria brevicollis (F.)   macropterous   601 records

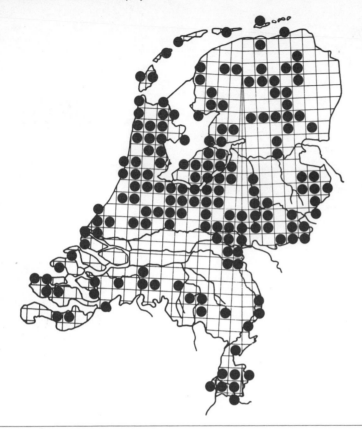

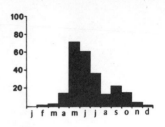

larvae

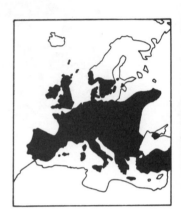

34 Nebria salina Fairm. *N. iberica Oliv.*   macropterous   99 records

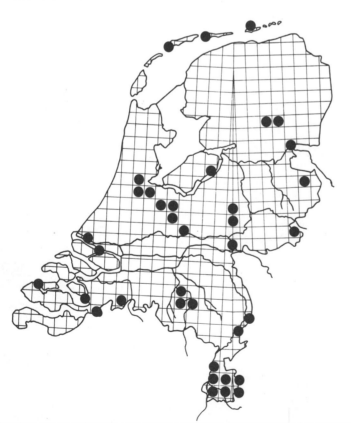

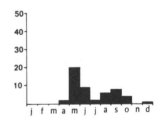

35 Notiophilus pusillus Wat. *N. aestuans Motsch.*

macropterous   47 records

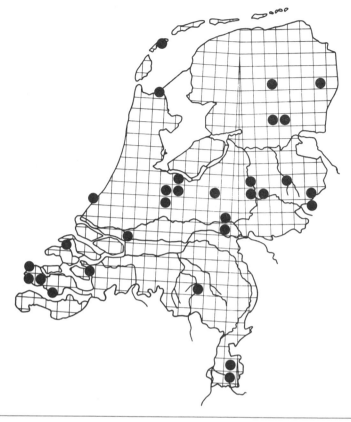

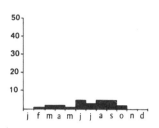

larvae

36 Notiophilus aquaticus (L.)

dimorphic   345 records

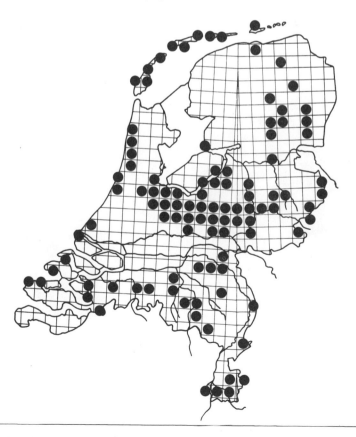

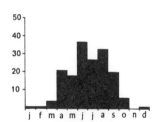

larvae

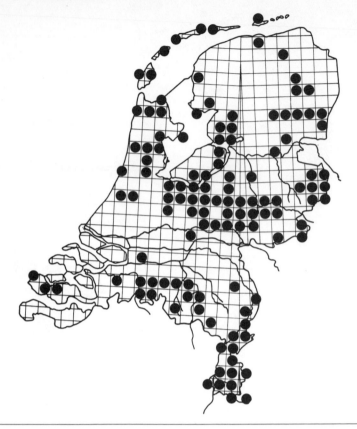
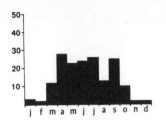

larvae

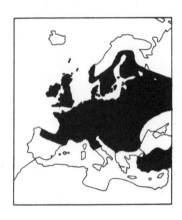

38   **Notiophilus hypocrita Curt.**   *N. germinyi Fauv.*   dimorphic   209 records

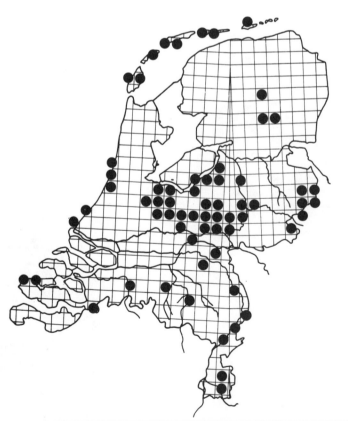
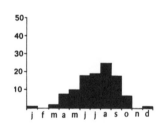

larvae

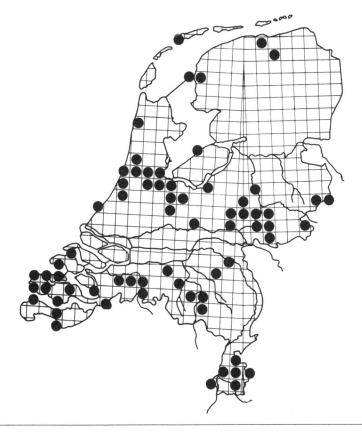

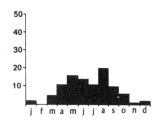

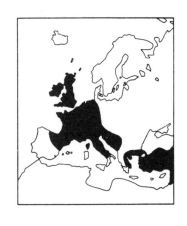

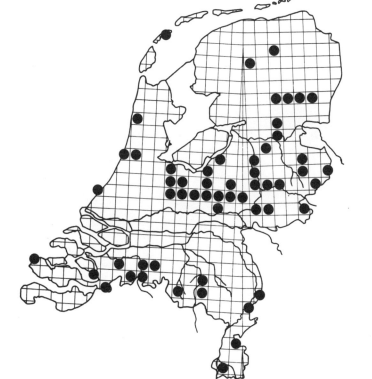

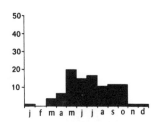

41 Notiophilus biguttatus (F.)

dimorphic   557 records

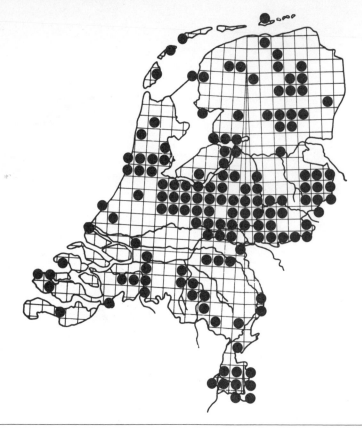

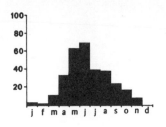

larvae

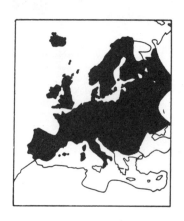

42 Notiophilus quadripunctatus Dej.

1 record

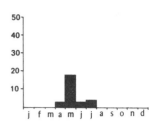

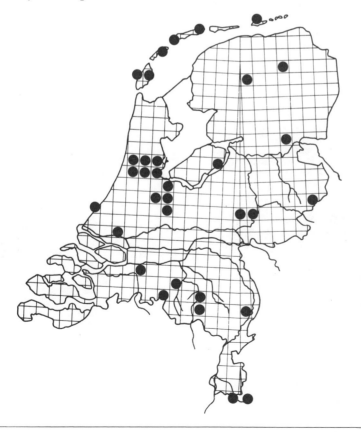

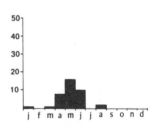

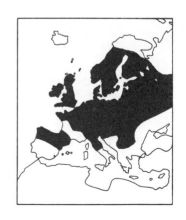

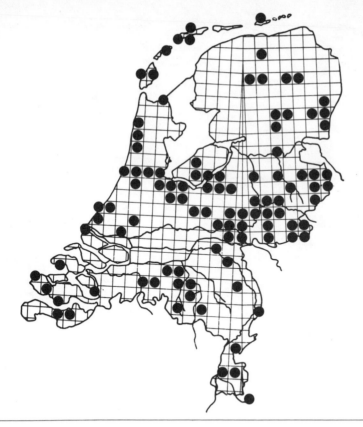

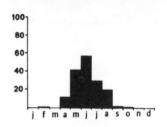

larvae

46 **Elaphrus riparius (L.)**   macropterous   335 records

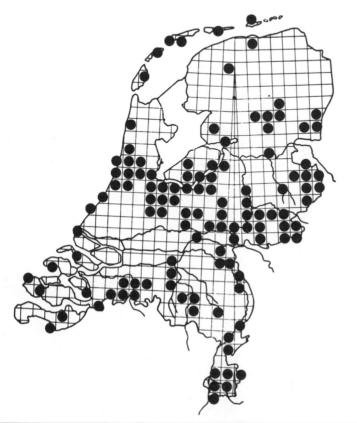

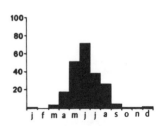

larvae

47 Elaphrus aureus Müll.

47 **Elaphrus aureus Müll.** 46 records

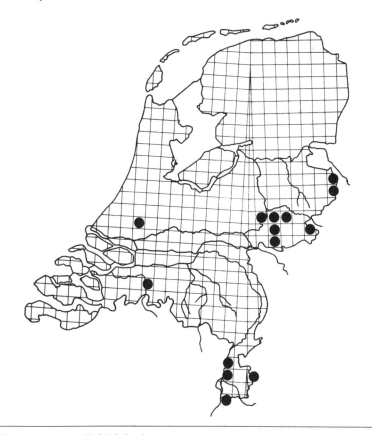

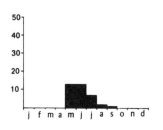

48 **Elaphrus ullrichi Redt.** 44 records

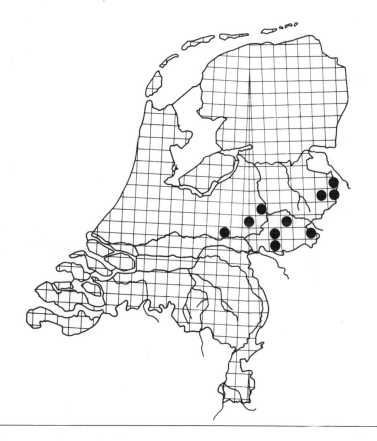

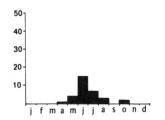

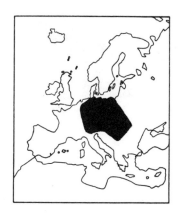

49   Loricera pilicornis (F.)

macropterous   663 records

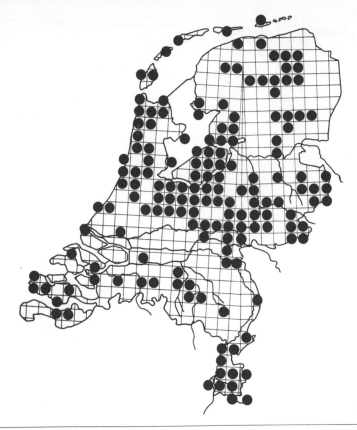
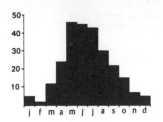

larvae

50   Clivina fossor (L.)

dimorphic   503 records

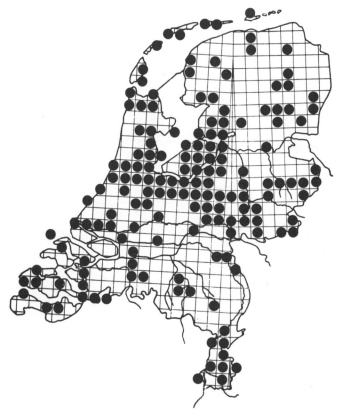
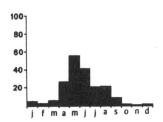

larvae

51　Clivina collaris (Hbst.)　*C. contracta (Fourcr.)*　　　macropterous　174 records

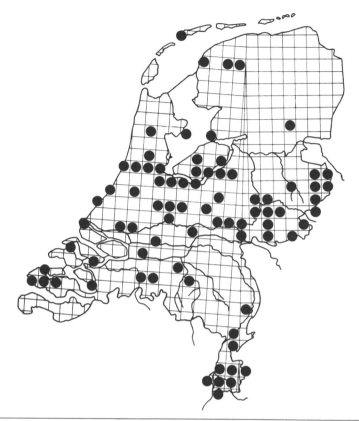
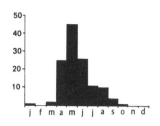

52　Dyschirius arenosus Steph.　*D. thoracicus (Rossi)*　　　macropterous　305 records

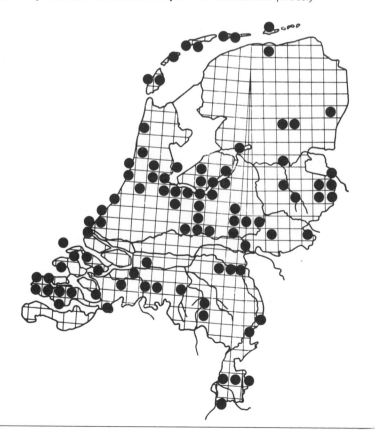
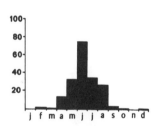

larvae

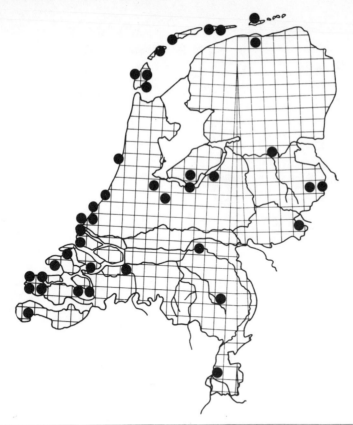

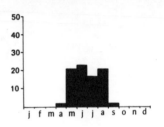

larvae

54 Dyschirius neresheimeri Wagn.

macropterous 1 record

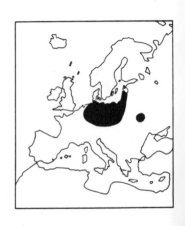

55 Dyschirius nitidus (Dej.)

macropterous 6 records

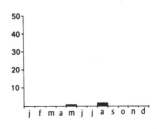

56 Dyschirius politus (Dej.)

macropterous 73 records

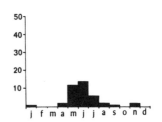

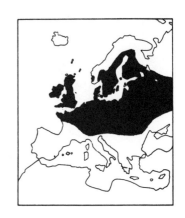

57 Dyschirius impunctipennis Daws.   macropterous 44 records

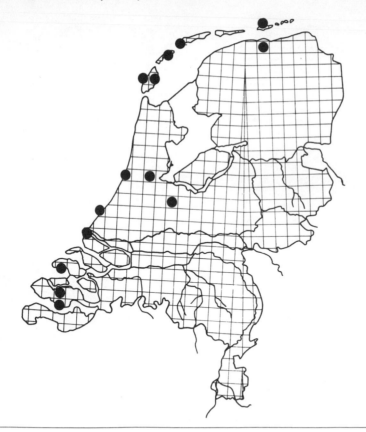
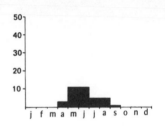
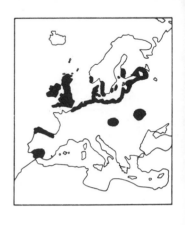

58 Dyschirius chalceus Er.   macropterous 12 records

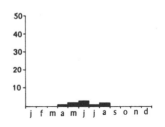

59 Dyschirius salinus Schaum.

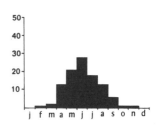

60 Dyschirius aeneus Dej.

macropterous   118 records

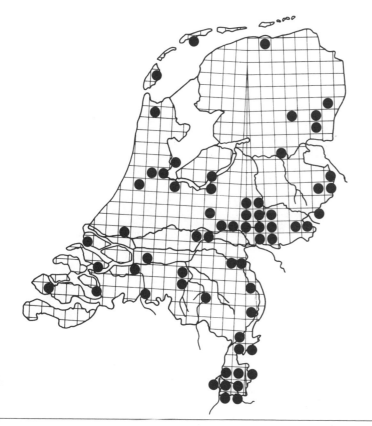
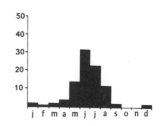

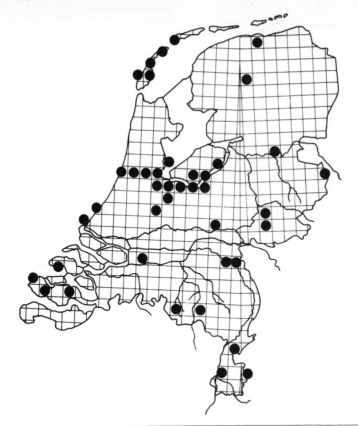

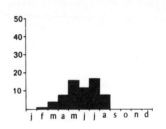

62   Dyschirius intermedius Putz.   macropterous   42 records

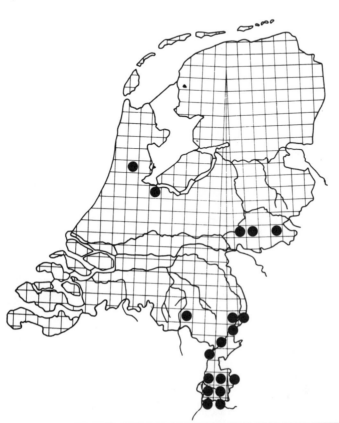

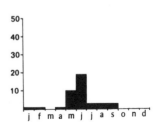

63   Dyschirius laeviusculus Putz.

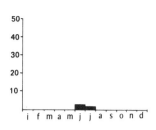

64   Dyschirius angustatus (Ahr.)

macropterous   15 records

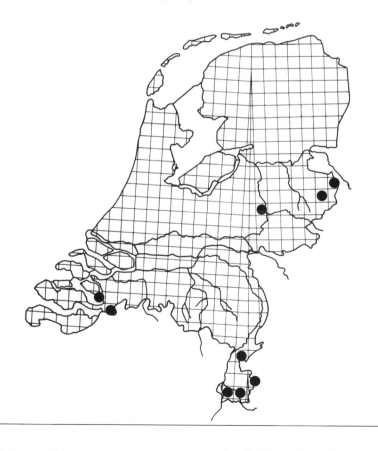
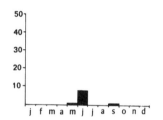

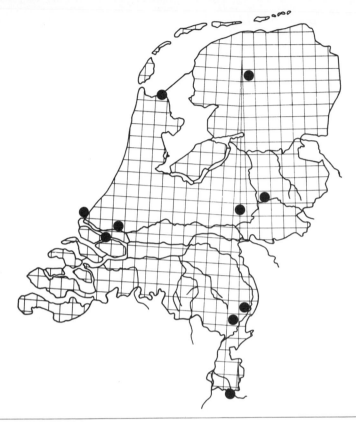

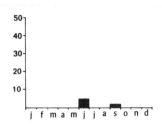

66 Dyschirius globosus (Hbst.)   dimorphic   611 records

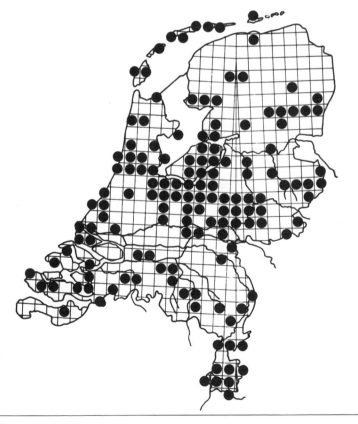

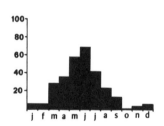

larvae

67 Omophron limbatum (F.)

macropterous 133 records

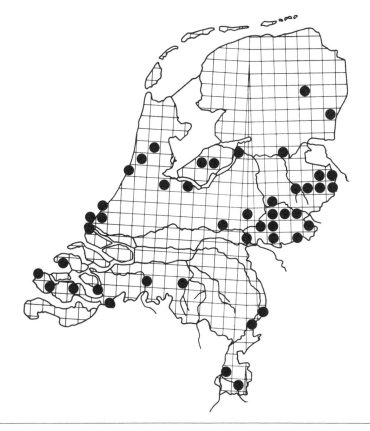

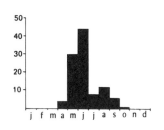

68 Broscus cephalotes (L.)

macropterous 352 records

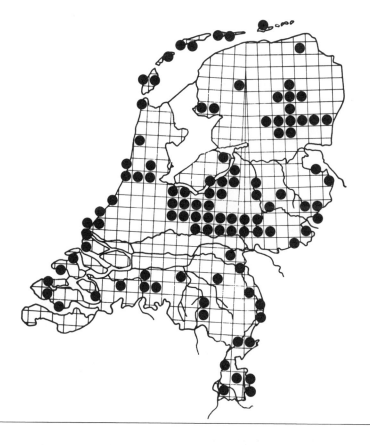

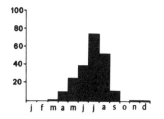

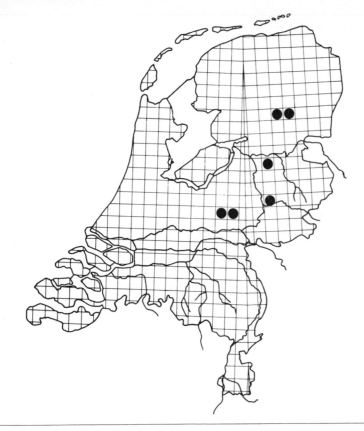

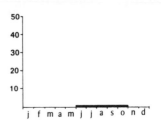

70 **Asaphidion pallipes (Dft.)** macropterous 103 records

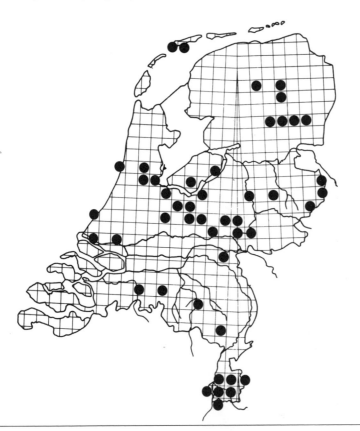

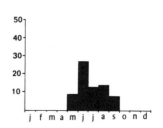

larvae

71 Asaphidion flavipes (L.)

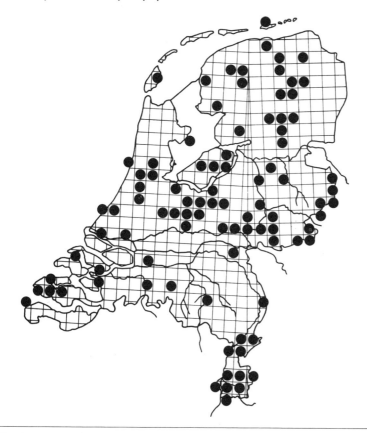

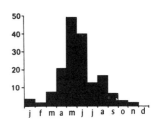

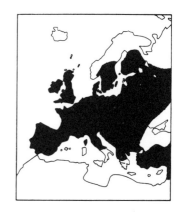

72 Bembidion striatum (F.)

macropterous   30 records

73 Bembidion argenteolum Ahr.

macropterous   95 records

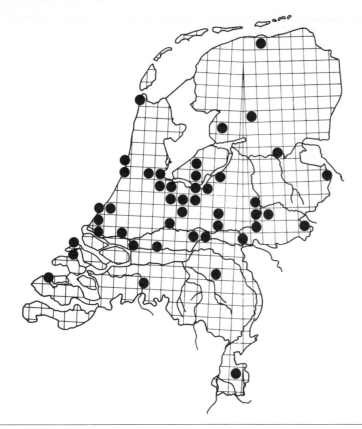

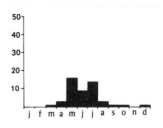

74 Bembidion velox (L.)

macropterous   56 records

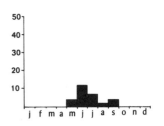

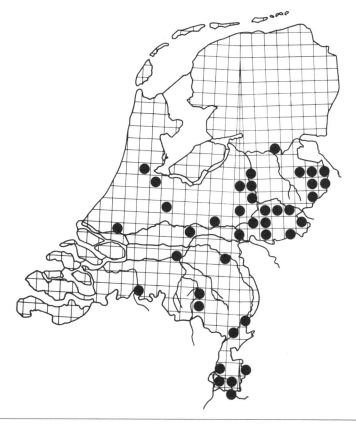

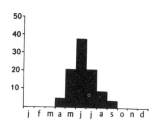

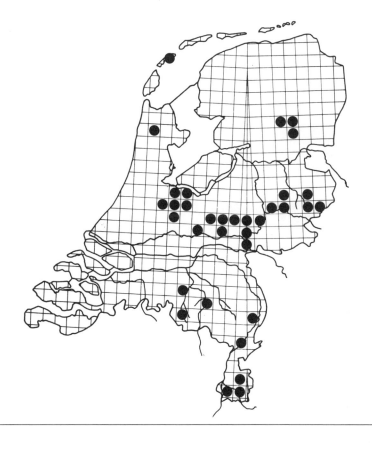

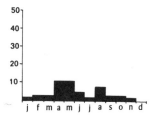

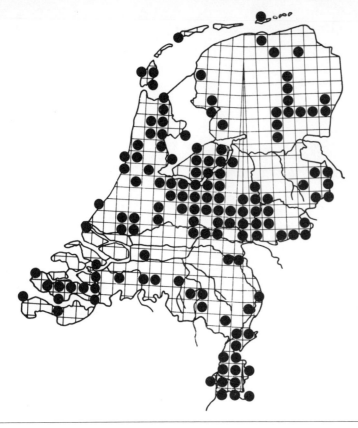

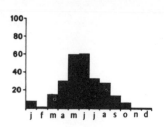

larvae

78   Bembidion properans Steph.   dimorphic   379 records

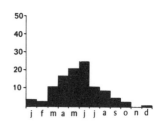

79 Bembidion punctulatum Drap.

macropterous   60 records

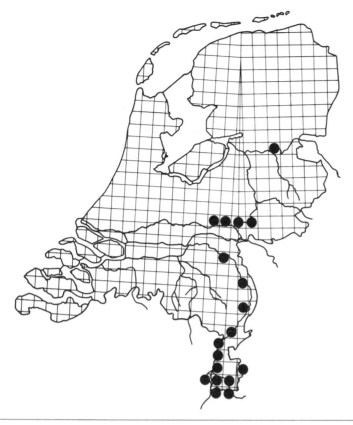

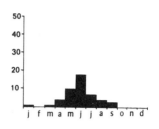

80 Bembidion pallidipenne Ill.

macropterous   103 records

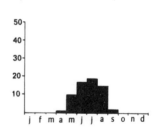

larvae

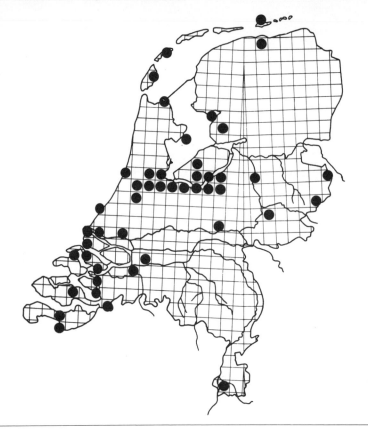
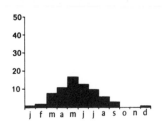

82   Bembidion dentellum (Thunb.)   macropterous   156 records

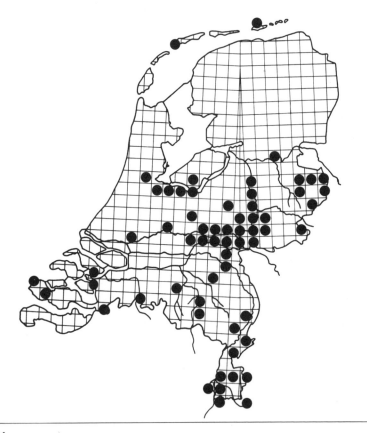
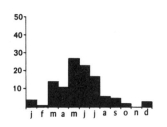

83 Bembidion varium (Oliv.) macropterous   264 records

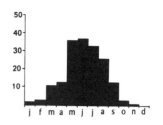

larvae

84 Bembidion obliquum Strm. macropterous   151 records

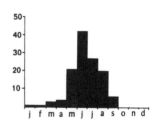

85 **Bembidion semipunctatum** Donov.   *B. adustum Schaum.*

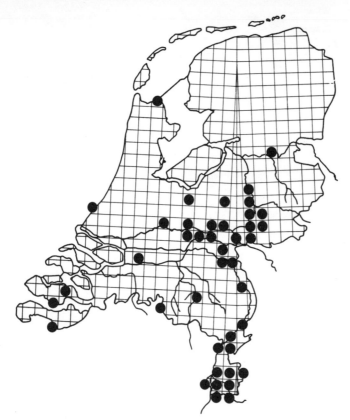
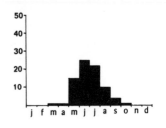
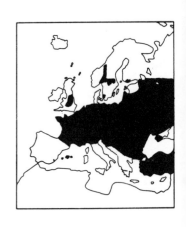

86 **Bembidion ephippium (Mrsh.)**

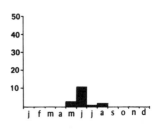
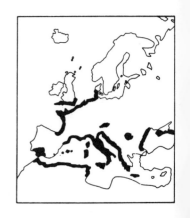

87   Bembidion prasinum (Dft.)   macropterous   5 records

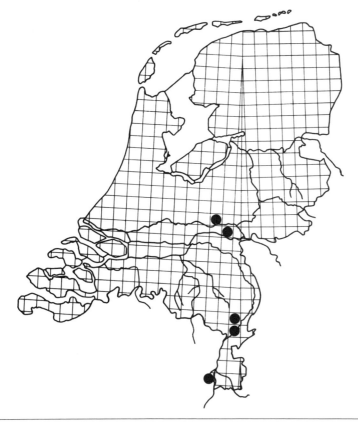

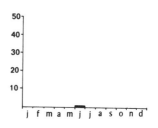

88   Bembidion tibiale (Dft.)   macropterous   29 records

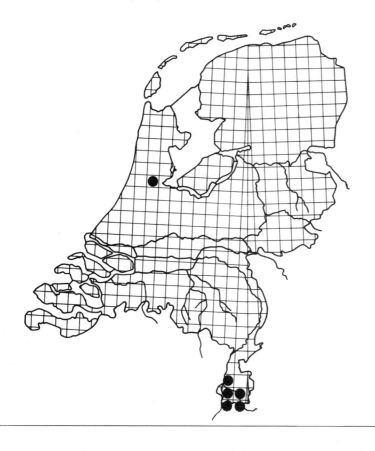

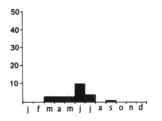

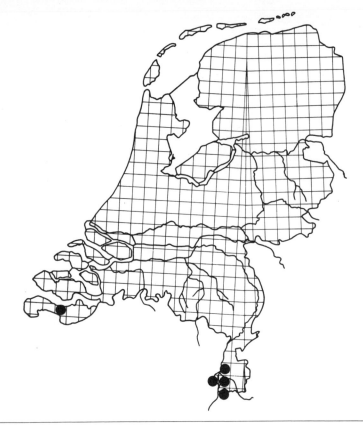

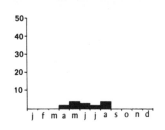

90 **Bembidion fasciolatum (Dft.)**   2 records

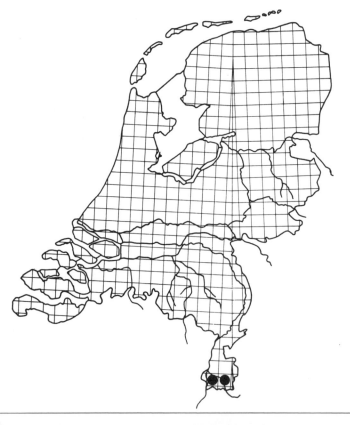

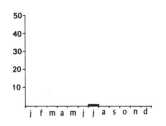

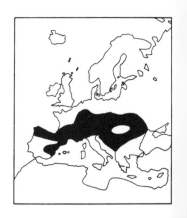

91 Bembicion monticola Strm.

macropterous   12 records

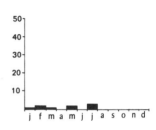

92 Bembidion nitidulum (Mrsh.)

macropterous   66 records

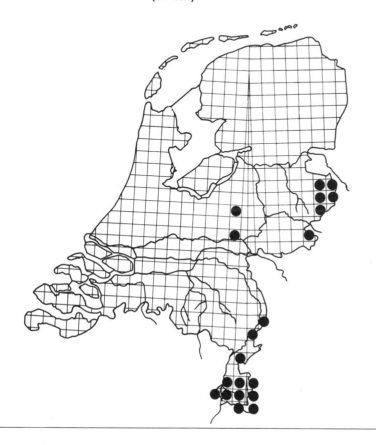

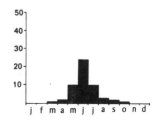

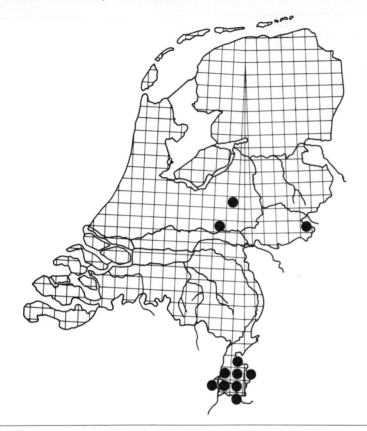

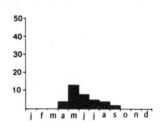

94   **Bembidion milleri Duv.**   13 records

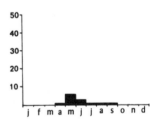

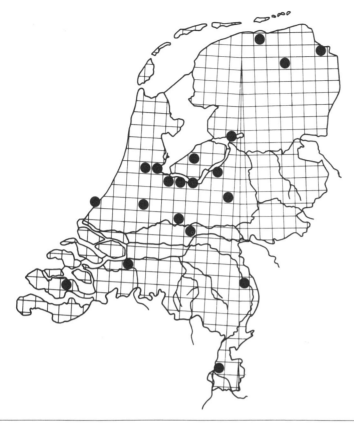

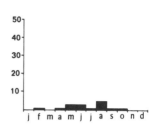

96   Bembidion rupestre (L.)   *B. bruxellense Wesmael*   macropterous   308 records

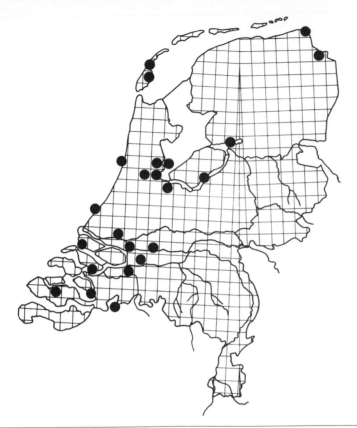

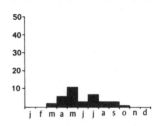

98 **Bembidion ustulatum (L.)** *B. tetracolum Say.*   dimorphic   474 records

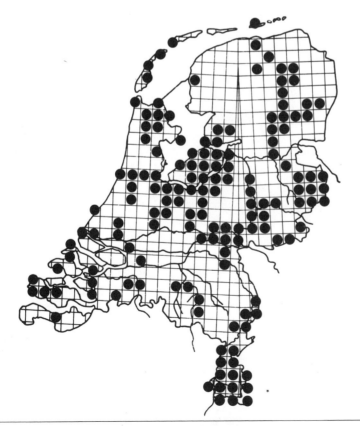

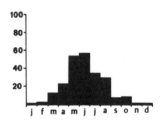

99 Bembidion femoratum Strm.

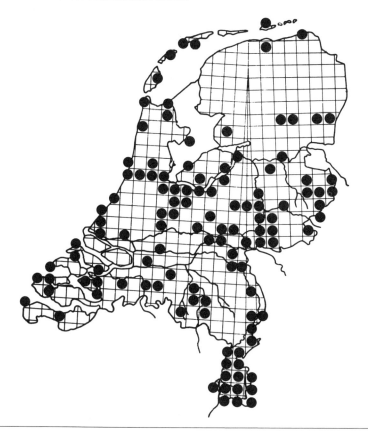
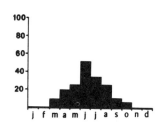

100 Bembidion testaceum (Dft.)

macropterous 44 records

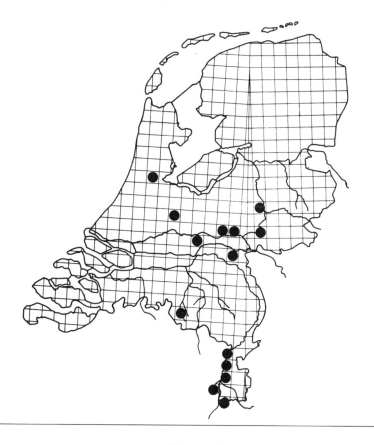
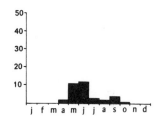

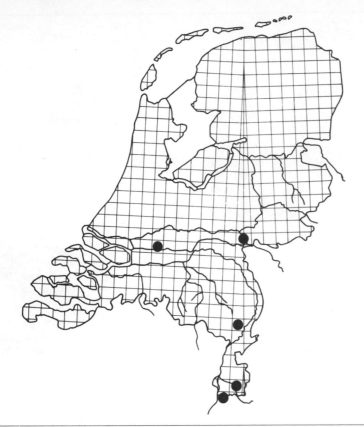

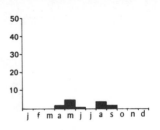

102   Bembidion decorum (Panz.)   macropterous   68 records

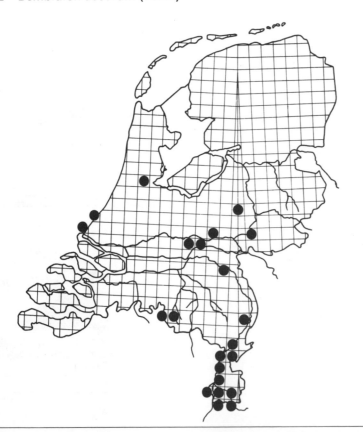

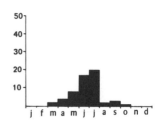

103   Bembidion modestum (F.)

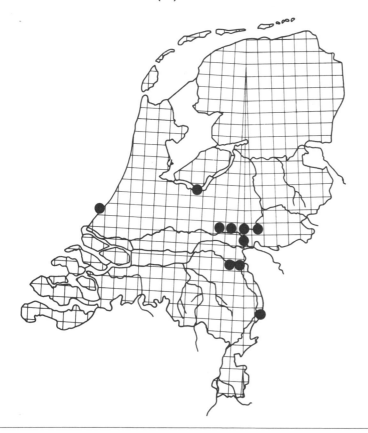

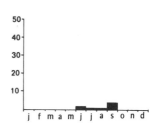

104   Bembidion illigeri Net.

macropterous   136 records

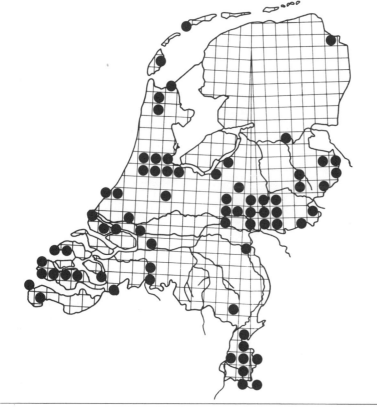

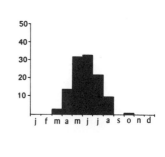

Bembidion stomoides Dej.   macropterous   50 records

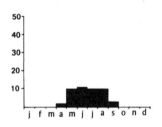

---

106   Bembidion elongatum Dej.   37 records

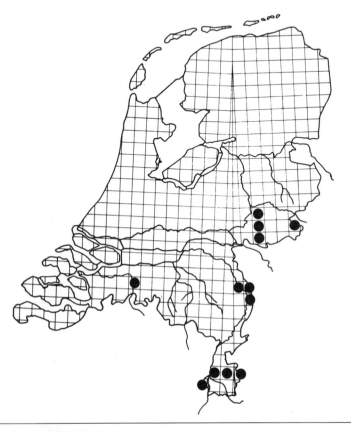
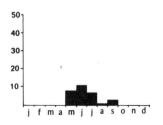

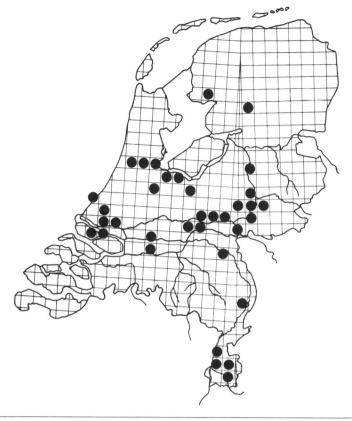

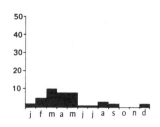

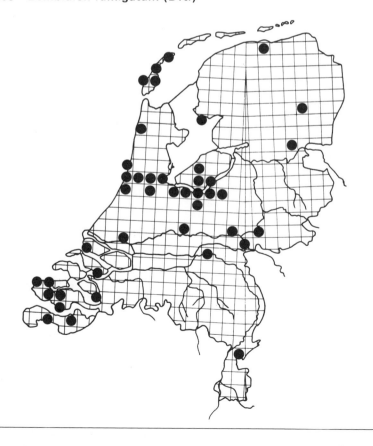

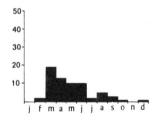

109 Bembidion assimile Gyll.

dimorphic   278 records

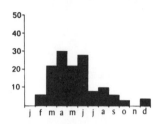

110 Bembidion normannum Dej.

macropterous   87 records

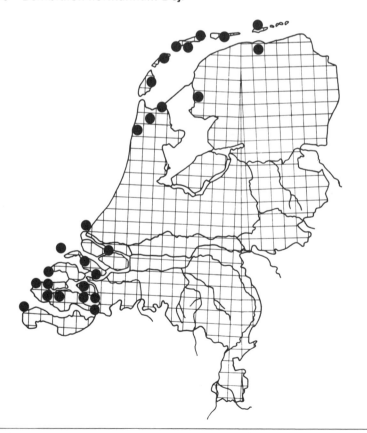

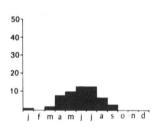

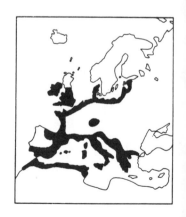

111   Bembidion minimum (F.)

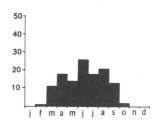

112   Bembidion tenellum Er.

macropterous   12 records

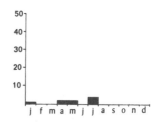

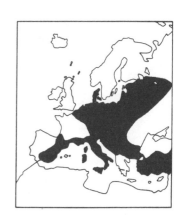

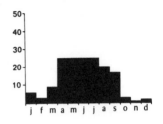

114   **Bembidion quadrimaculatum (L.)**   macropterous 279 records

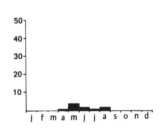

115 Bembidion quadripustulatum Serv. *B. quadriguttatum (Oliv.)*   macropterous   75 records

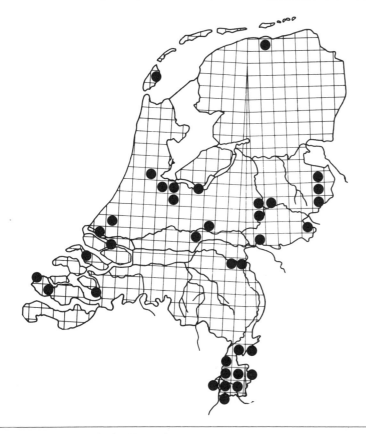

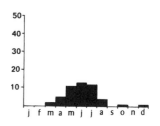

116 Bembidion doris (Panz.)   macropterous   116 records

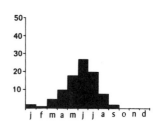

117 Bembidion articulatum (Panz.)

macropterous   150 records

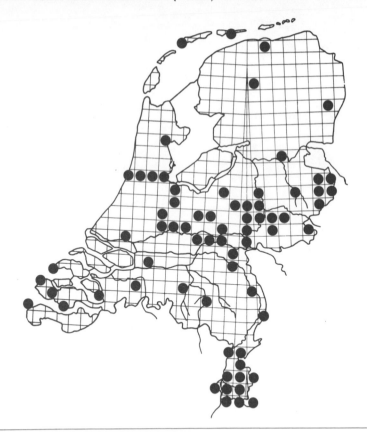

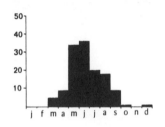

118 Bembidion octomaculatum Goeze

macropterous   14 records

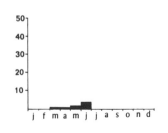

119 Bembidion obtusum Serv.

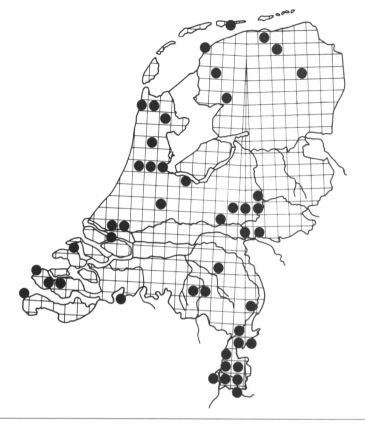

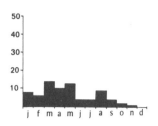

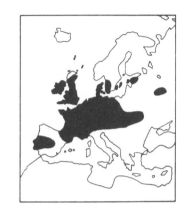

120 Bembidion biguttatum (F.)

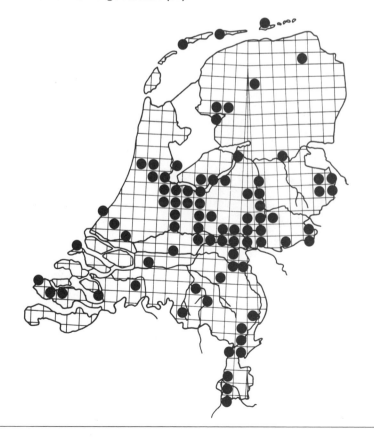

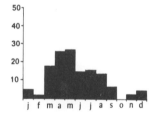

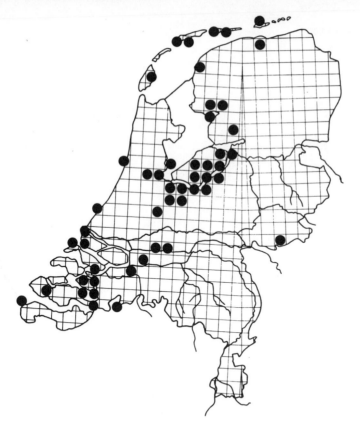

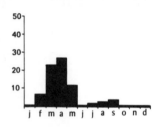

122 Bembidion unicolor Chd. *B. mannerheimi auct.*   brachypterous   79 records

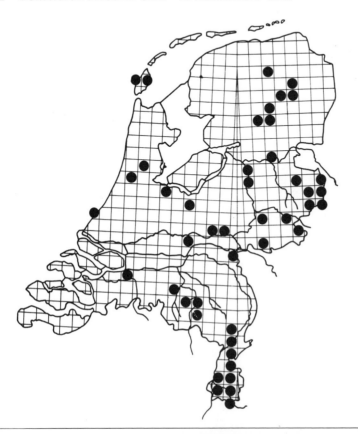

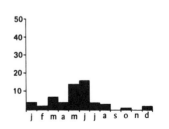

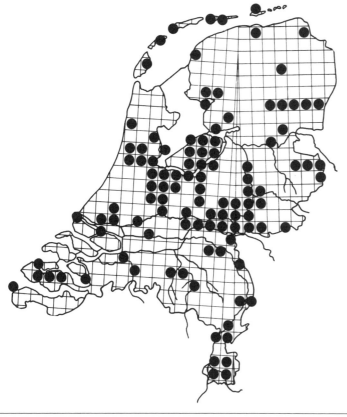

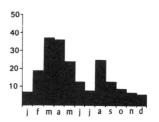

larvae

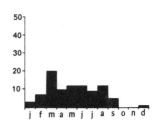

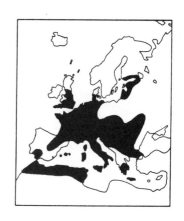

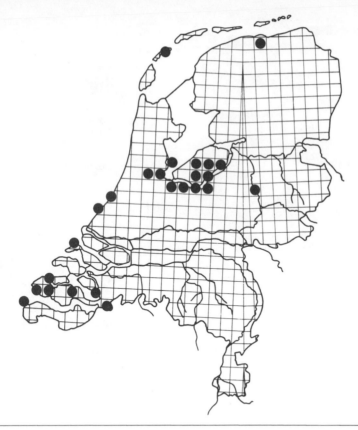

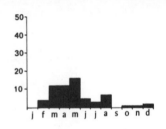

126 Cillenus lateralis Sam. *Bembidion laterale Sam*   dimorphic 46 records

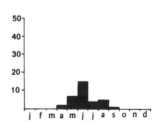

larvae

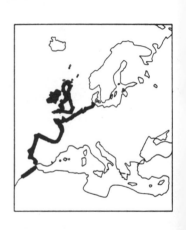

127 **Ocys harpaloides Serv.** *Bembidion harpaloides Serv.* macropterous 65 records

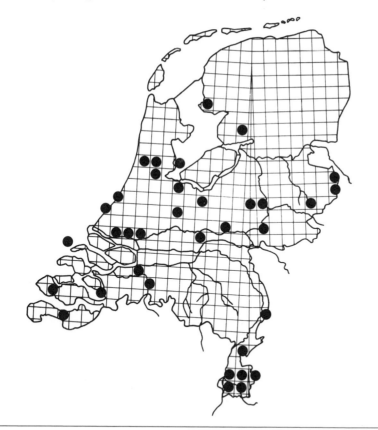
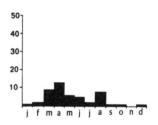

128 **Ocys quinquestriatus (Gyll.)** *Bembidion quinquestriatum (Gyll.)* macropterous 48 records

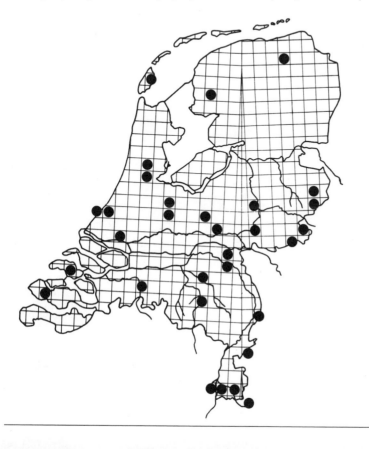
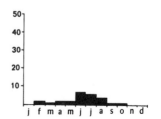

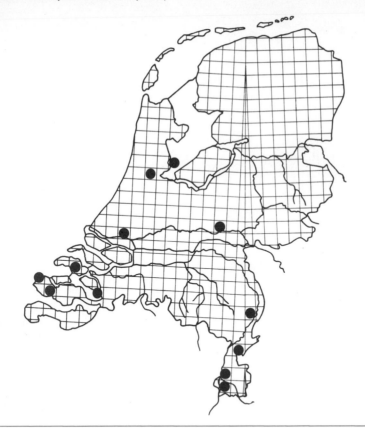
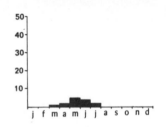

130 Tachys micros Fisch.-W.   *T. gregarius Chd.*   41 records

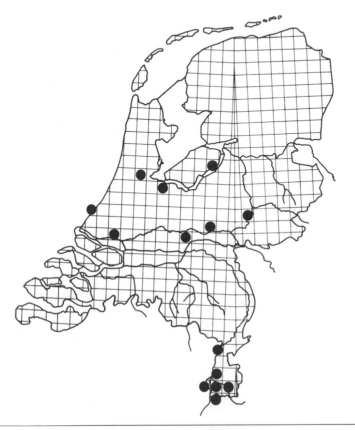
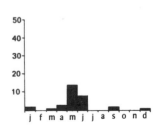

131   Tachys scutellaris (Germ.)

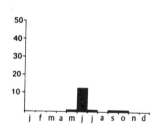

132   Tachys parvulus (Dej.)

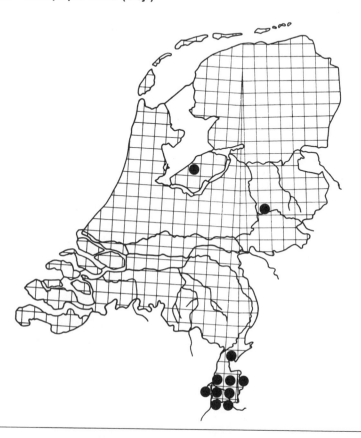

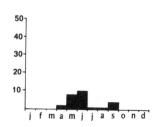

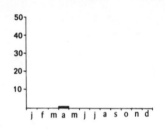

134 Tachys bisulcatus (Nic.)   macropterous 11 records

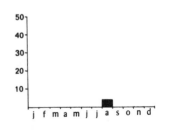

135 Tachyta nana (Gyll.)

macropterous   2 records

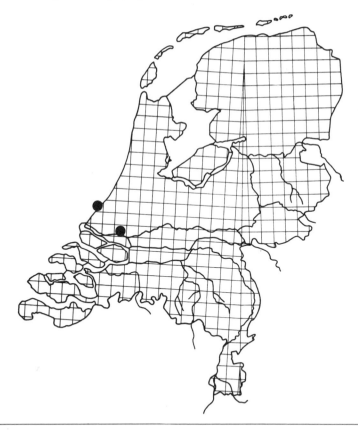

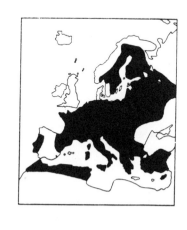

---

136 Perileptus areolatus (Creutz.)

macropterous   8 records

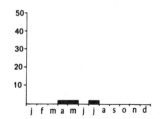

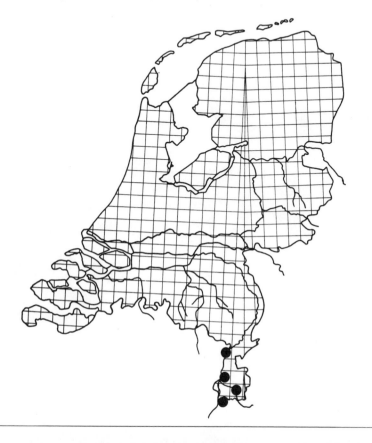

137   Trechus secalis (Payk.)

brachypterous   65 records

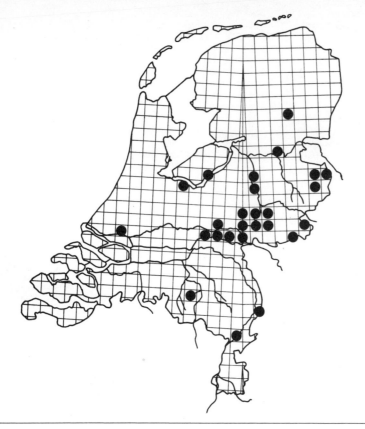

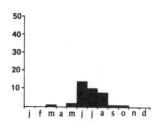

138   Trechus rubens (F.)

macropterous   37 records

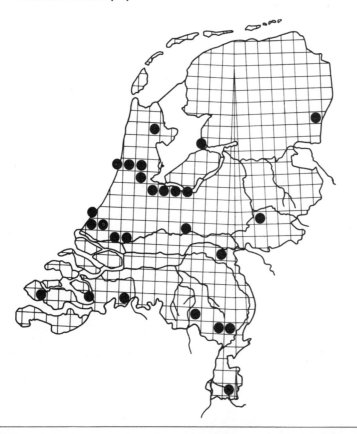

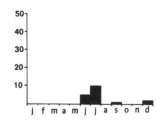

larvae

139 Trechus quadristriatus (Schrk.)

macropterous   406 records

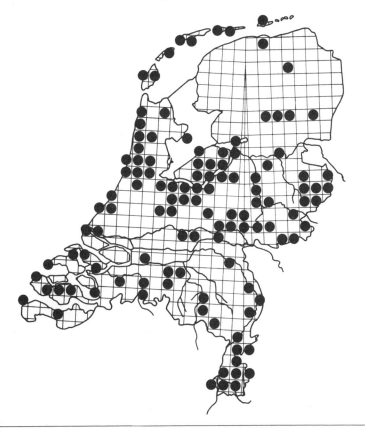

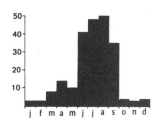

larvae

140 Trechus obtusus Er.

dimorphic   262 records

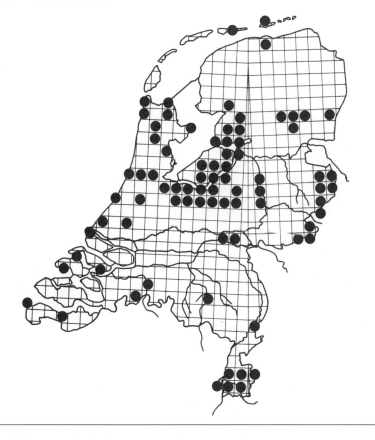

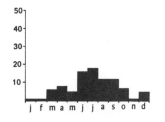

141 Trechus micros (Hbst.)

75 records

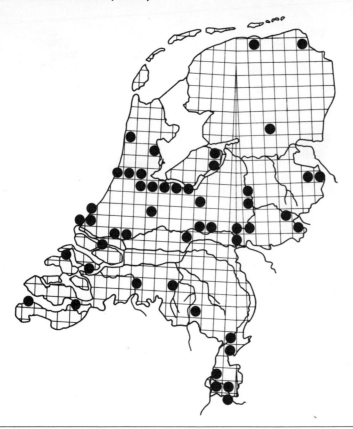
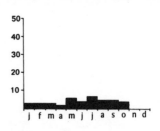

142 Trechus discus (F.)

macropterous   142 records

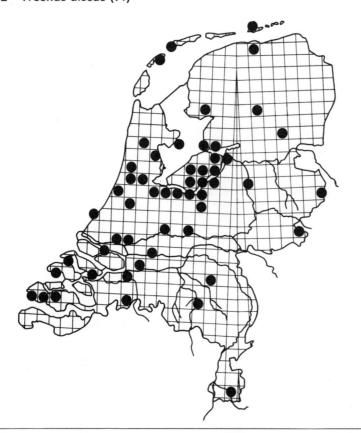
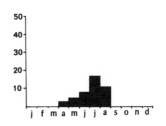

143 Pogonus luridipennis (Germ.)

macropterous   23 records

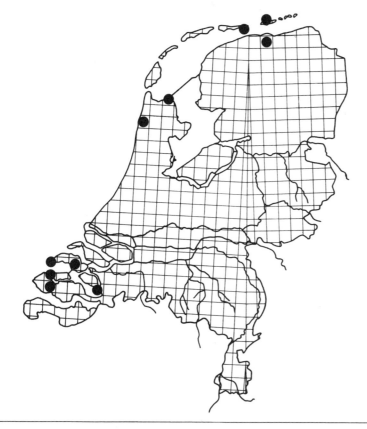

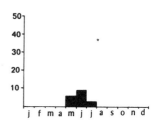

144 Pogonus chalceus (Mrsh.)

macropterous   141 records

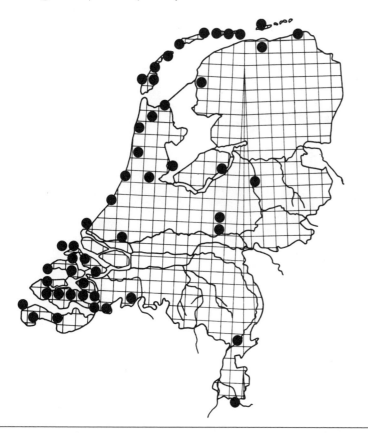

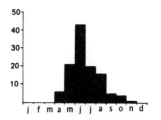

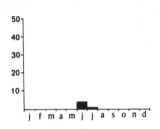

146 **Patrobus atrorufus (Ström.)** *P. excavatus (Payk.)*   brachypterous   155 records

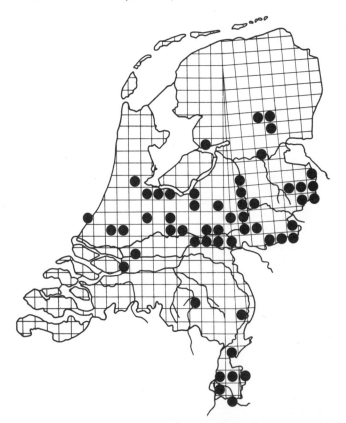
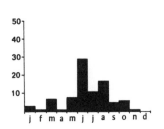
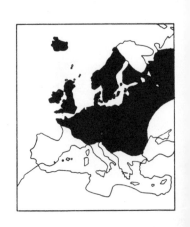

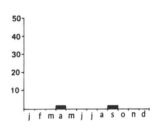

148   **Panagaeus cruxmajor (L.)**   macropterous   139 records

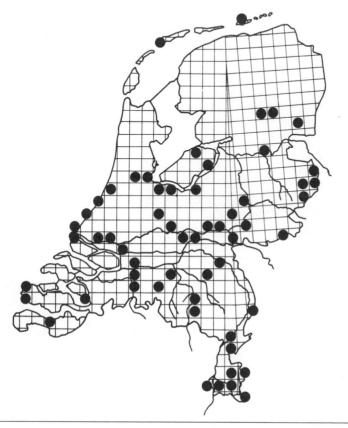

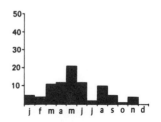

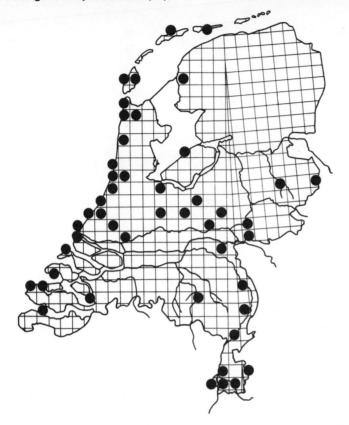

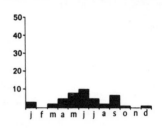

150 **Callistus lunatus (F.)** 53 records

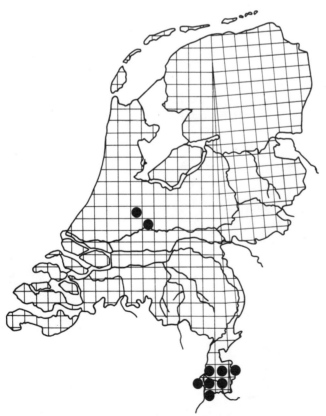

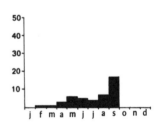

151 Chlaenius tristis (Schall.) macropterous 5 records

152 Chlaenius nigricornis (F.) macropterous 175 records

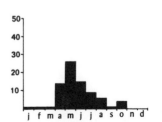

larvae

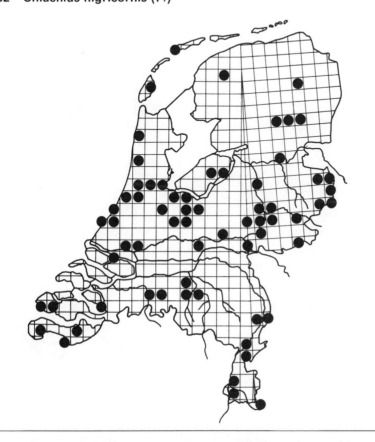

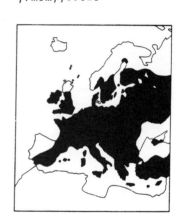

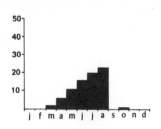

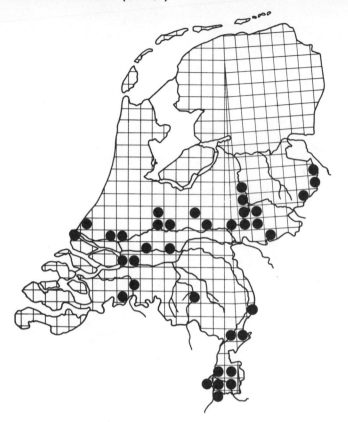

154   **Chlaenius vestitus (Payk.)**   macropterous   112 records

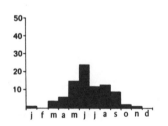

larvae

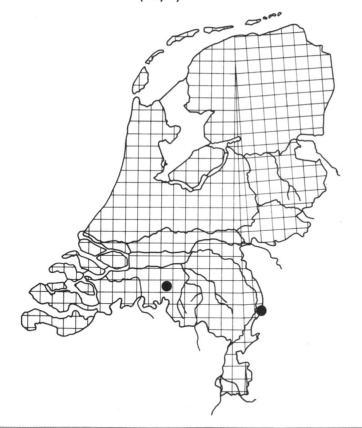

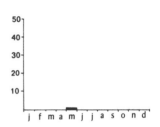

larvae

---

156 Oodes helopioides (F.)

macropterous   142 records

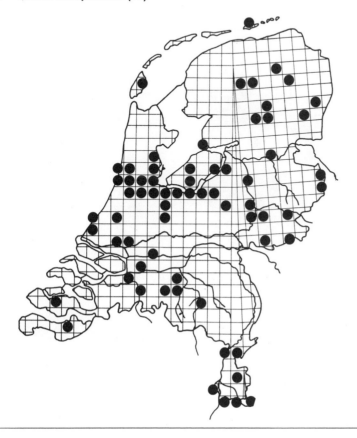

macropterous   31 records

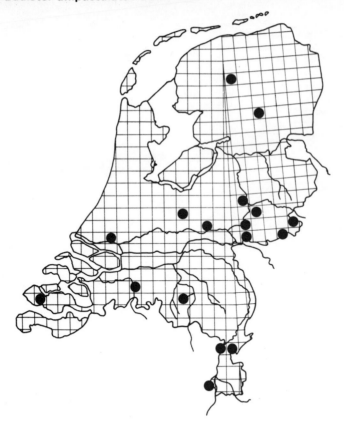

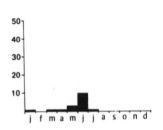

macropterus   265 records

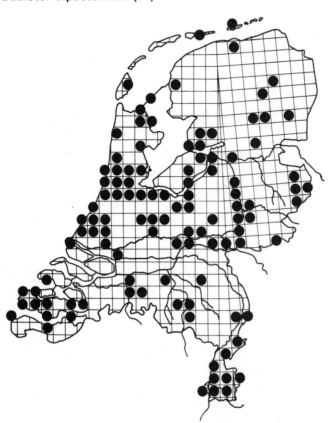

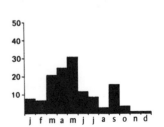

larvae

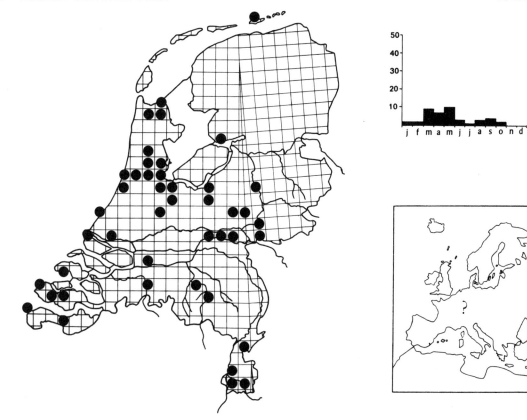

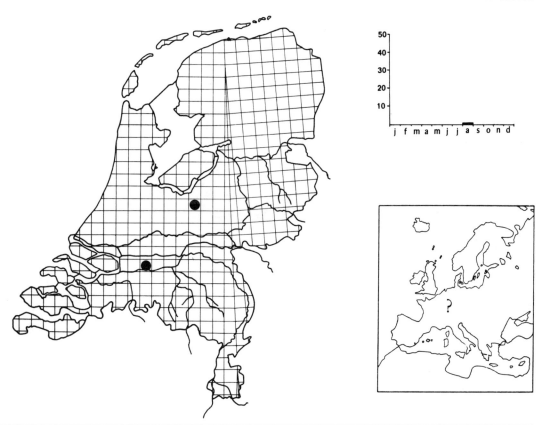

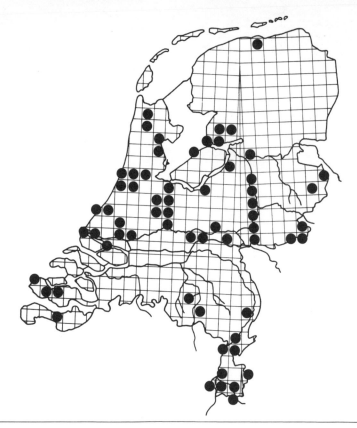

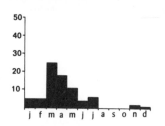

162 Badister peltatus (Panz.)   macropterous 46 records

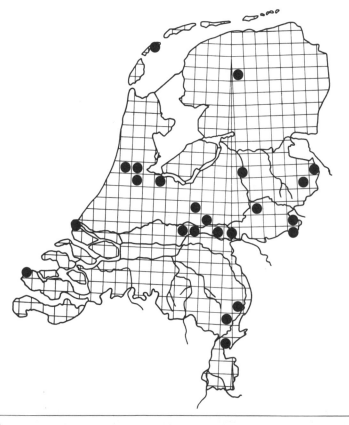

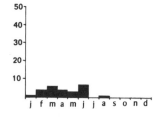

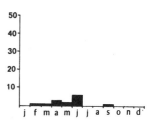

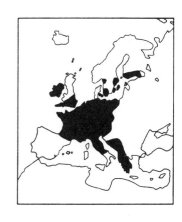

164 Badister anomalus (Perr.)   *B. striatulus Haus.*   9 records

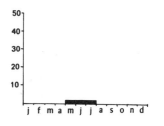

larvae
j f m a m j j a s o n d

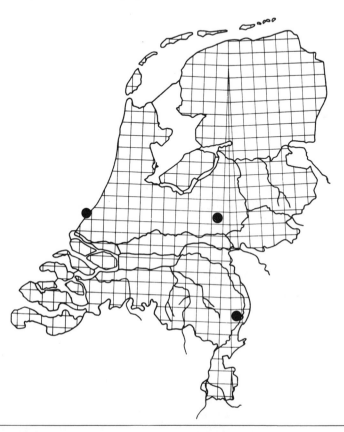

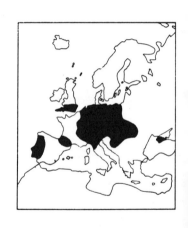

167 Harpalus diffinis Dej.

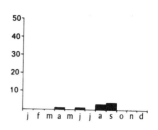

168 Harpalus punctatulus (Dft.)

macropterous   23 records

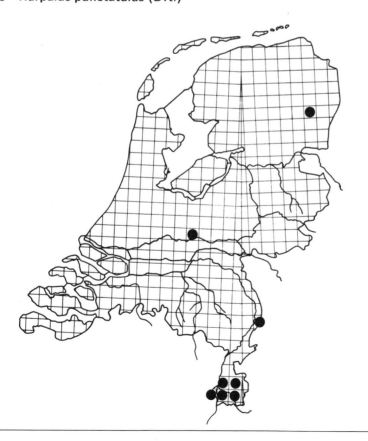

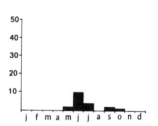

macropterous
larvae

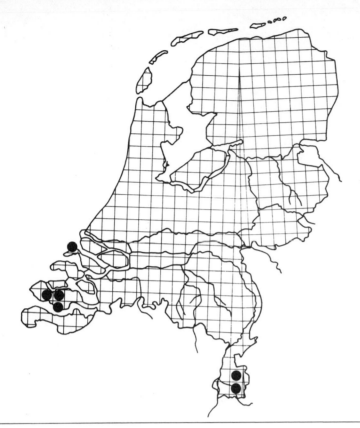
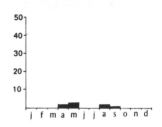

170 **Harpalus brevicollis** Serv.   *H. rufibarbis F., seladon Schaub.*   72 records

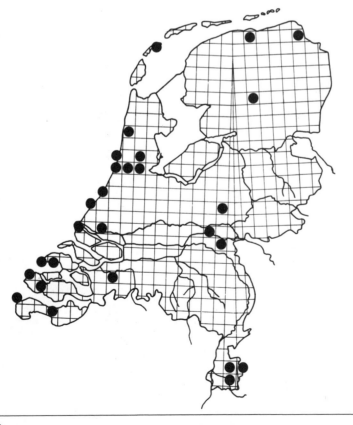
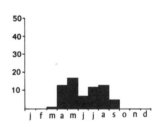
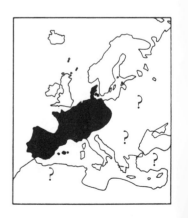

171 Harpalus rufibarbis Redt. *H. schaubergerianus Puei.* macropterous 22 records

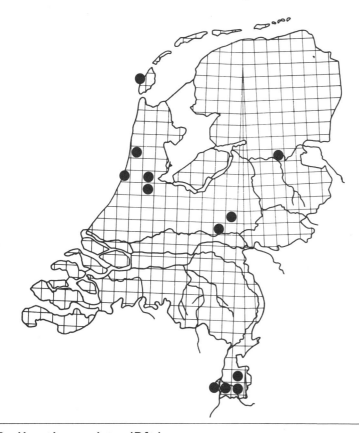

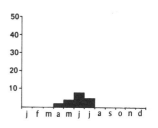

172 Harpalus cordatus (Dft.) 26 records

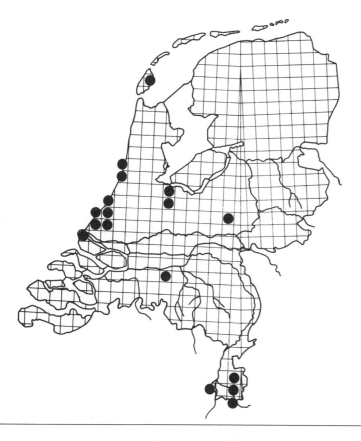

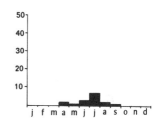

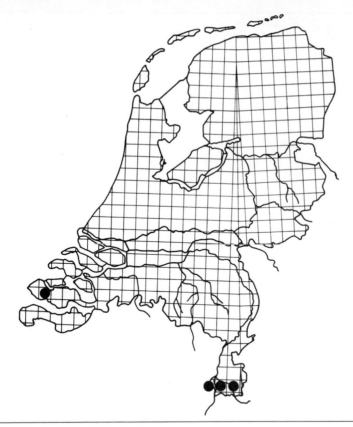

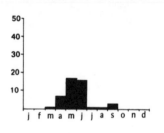

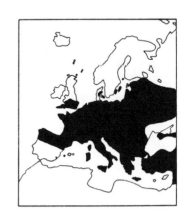

174 Harpalus punticollis (Payk.)   macropterous   16 records

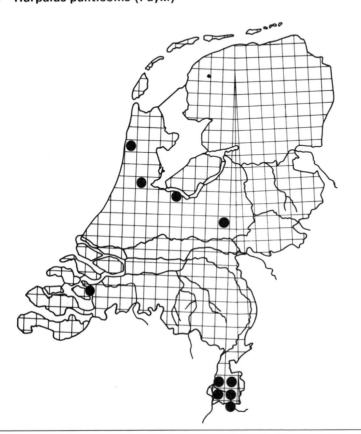

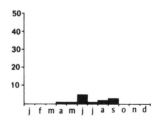

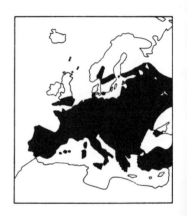

175   Harpalus melleti Heer   *H. brevicollis Serv. sensu Jeannel, H. rectangulus Thoms.*   22 records

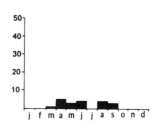

176   Harpalus zigzag Costa   *H. parallelus Schaum.*   7 records

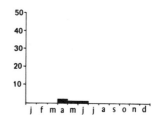

177 **Harpalus puncticeps Steph.** *H. angusticolle Müll, H. rectangulus Sharp* macropterous 51 records

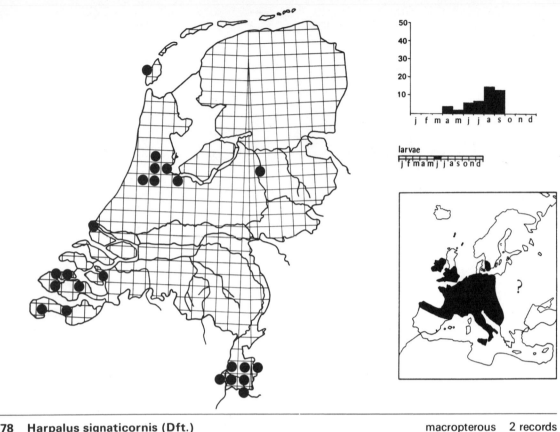

178 **Harpalus signaticornis (Dft.)** macropterous 2 records

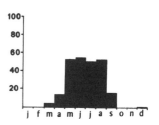

larvae

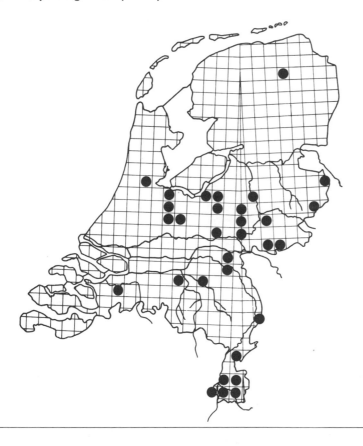

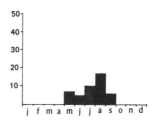

181 Harpalus calceatus (Dft.) macropterous   21 records

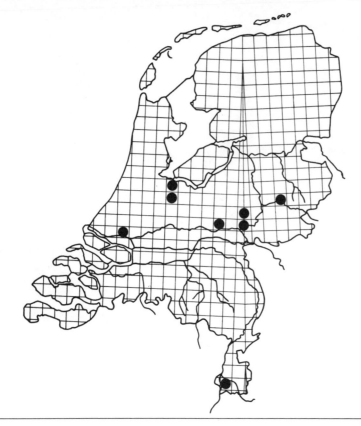

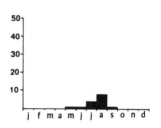

182 Harpalus rufus Brugg. macropterous   13 records

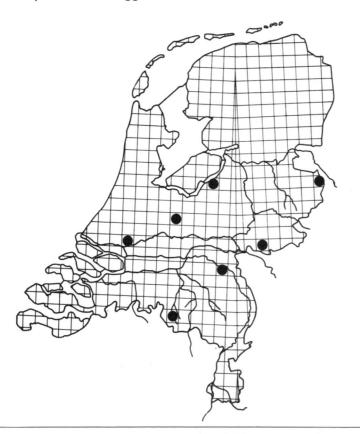

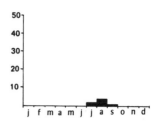

183 Harpalus froehlichi Strm.   macropterous   55 records

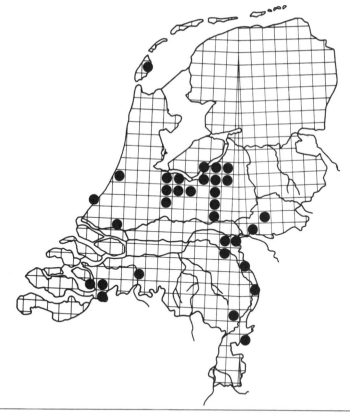

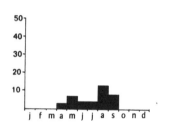

184 Harpalus aeneus (F.)   macropterous   672 records

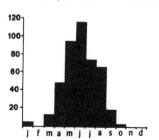

larvae

185 Harpalus distinguendus (Dft.)

macropterous 105 records

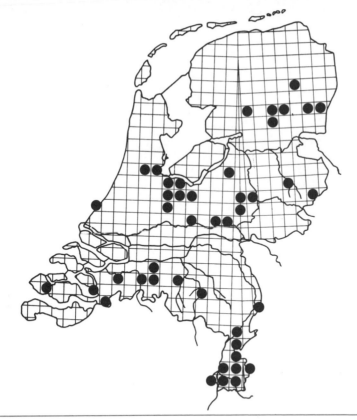
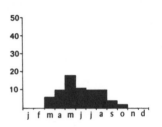

186 Harpalus smaragdinus (Dft.)

macropterous 102 records

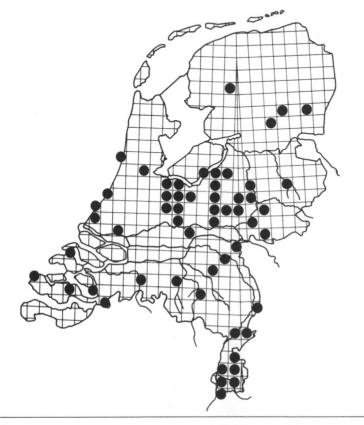

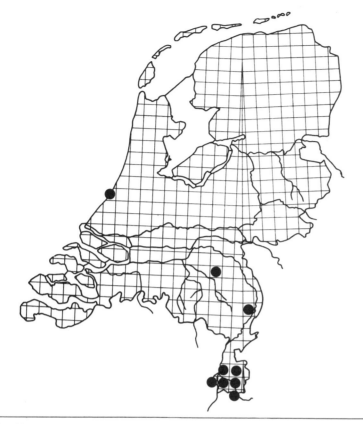

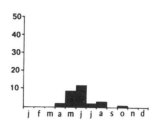

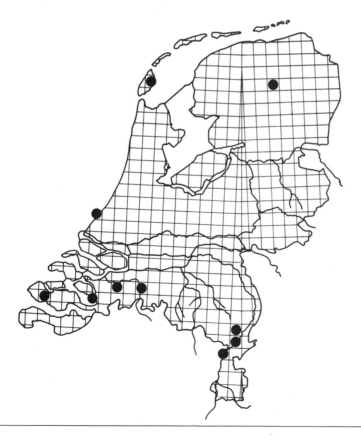

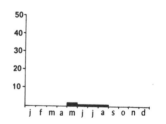

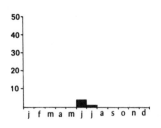

190 **Harpalus fuliginosus (Dft.)** macropterous 80 records

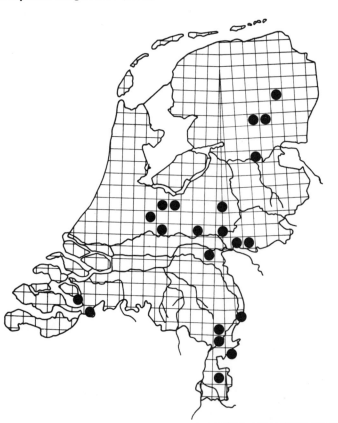
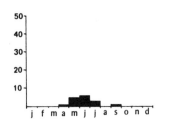

191 Harpalus winkleri Schaub.

macropterous 19 records

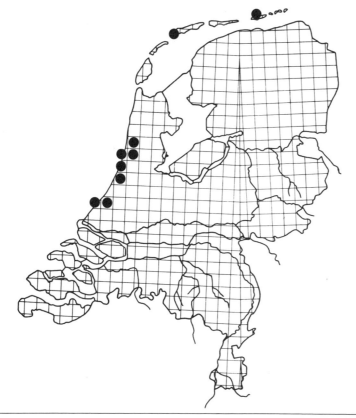

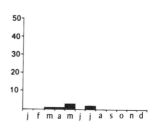

192 Harpalus latus (L.)

macropterous 243 records

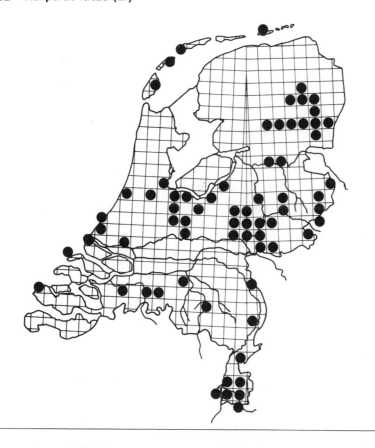

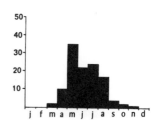

larvae

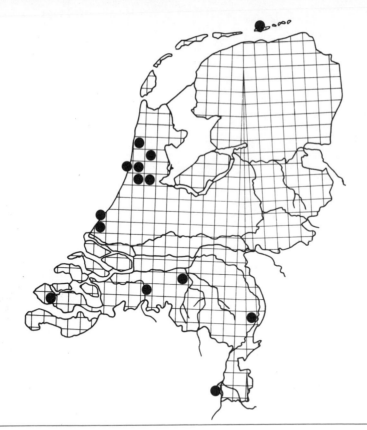
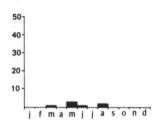

194   Harpalus quadripunctatus Dej.   macropterous   97 records

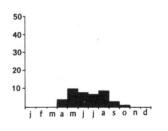

195 Harpalus rubripes (Dft.)   macropterous 183 records

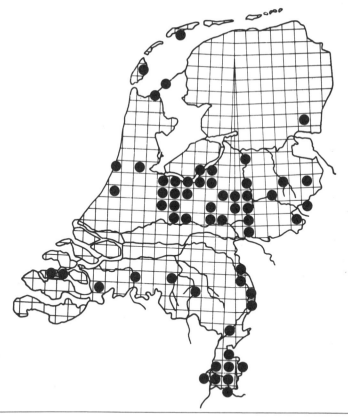
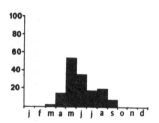

196 Harpalus honestus (Dft.)   23 records

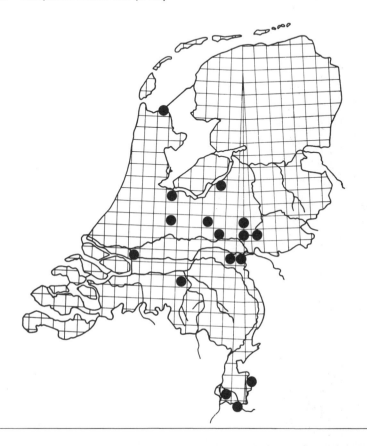

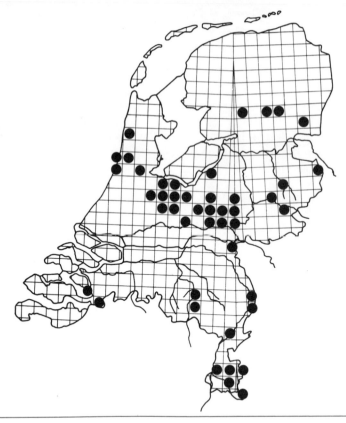

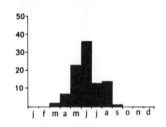

198 Harpalus neglectus Serv.   *H. neglectus Dej.*   dimorphic   100 records

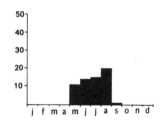

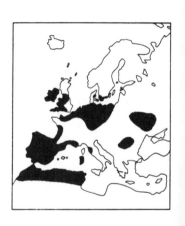

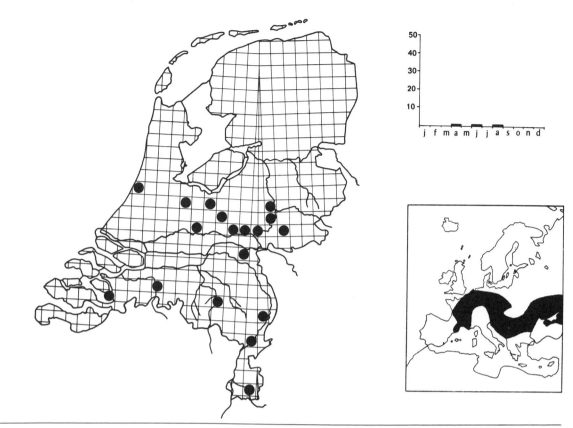

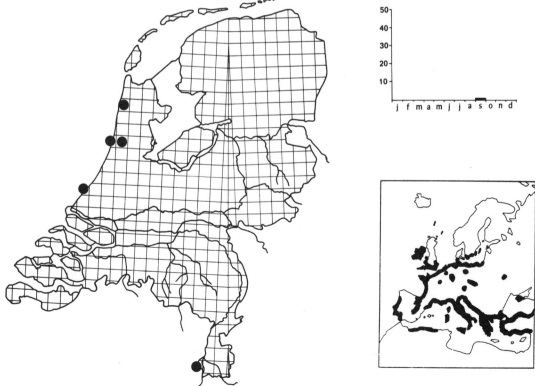

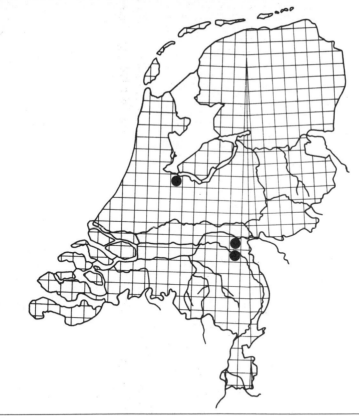

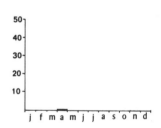

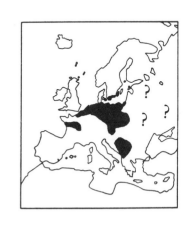

202   Harpalus vernalis (Dft.)   brachypterous   91 records

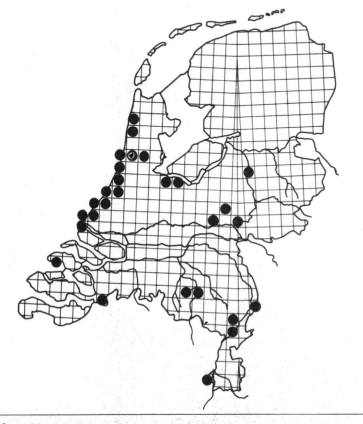

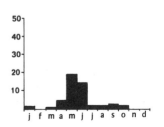

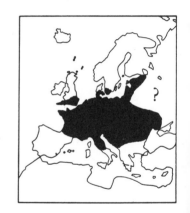

203 Harpalus servus (Dft.)

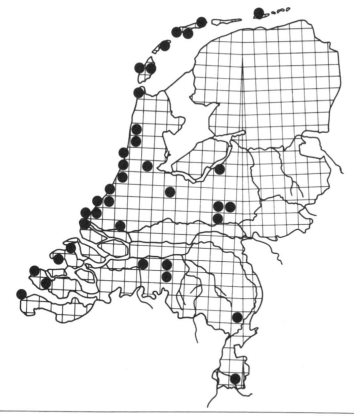

204 Harpalus tardus (Panz.)

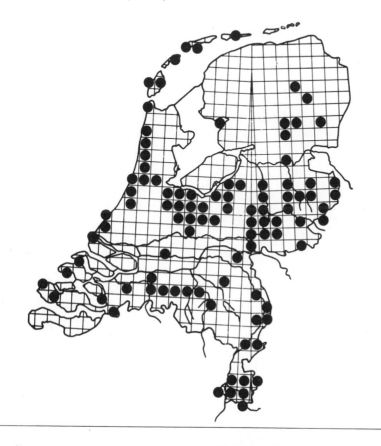

larvae

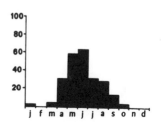

205 Harpalus modestus Dej.

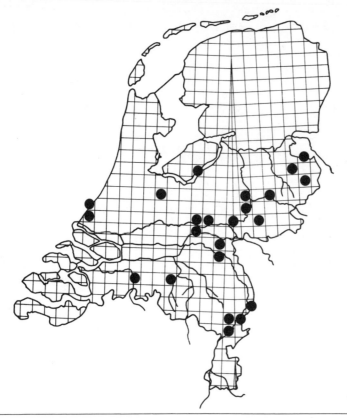

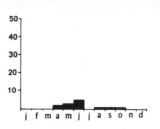

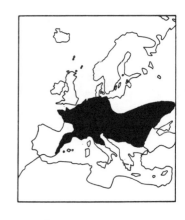

206 Harpalus anxius (Dft.)

macropterous   150 records

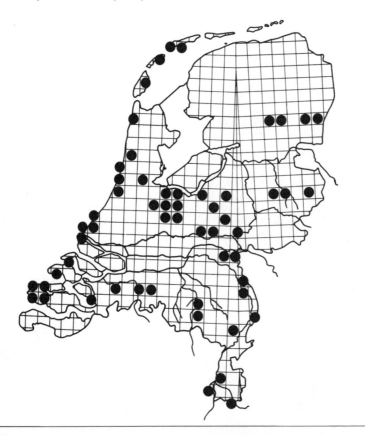

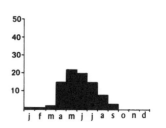

macropterous 59 records

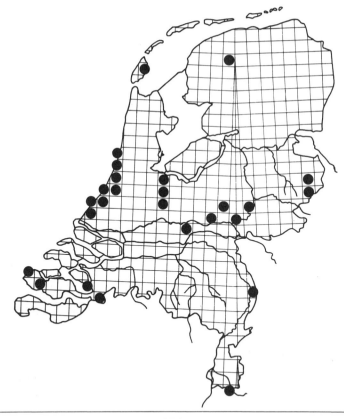

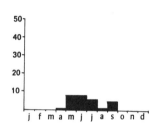

208 Parophonus maculicornis (Dft.)

16 records

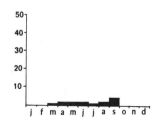

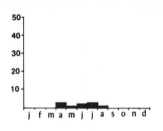

210 Trichotichnus nitens Heer

20 records

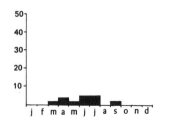

211   Stenolophus teutonus (Schrk.)   macropterous   152 records

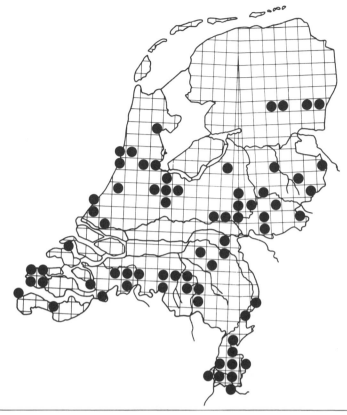

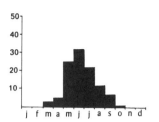

larvae

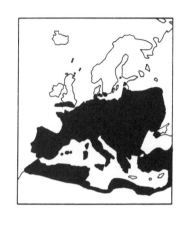

212   Stenolophus skrimshiranus Steph.   macropterous   54 records

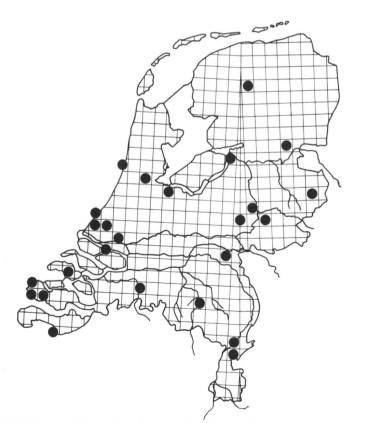

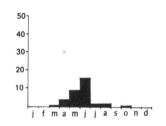

213 Stenolophus mixtus Hbst.

macropterous 348 records

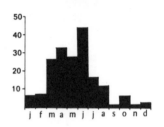

214 Acupalpus elegans Dej.

21 records

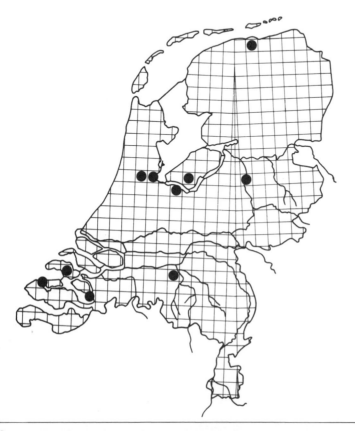

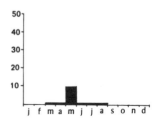

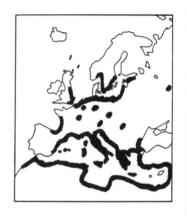

215 Acupalpus flavicollis (Strm.)

macropterous   165 records

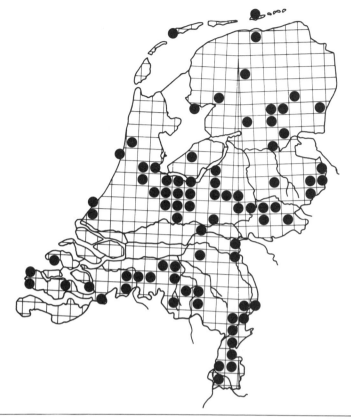

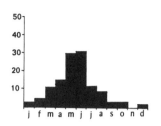

216 Acupalpus brunnipes (Strm.)

macropterous   54 records

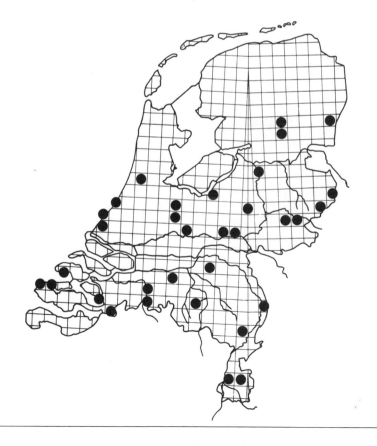

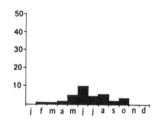

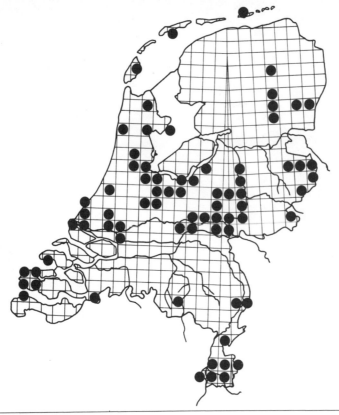

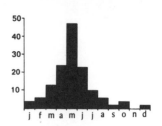

larvae

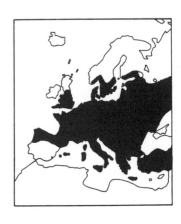

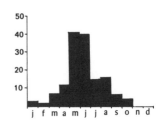

219 Acupalpus dubius Schilsky *A. luridus auct.*

macropterous   88 records

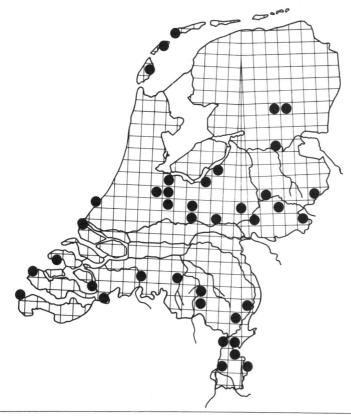
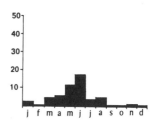

220 Acupalpus exiguus Dej.

macropterous   81 records

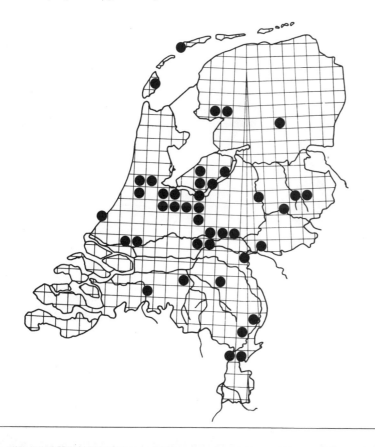
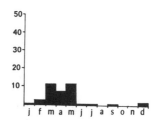

221 Acupalpus consputus (Dft.)

macropterous   55 records

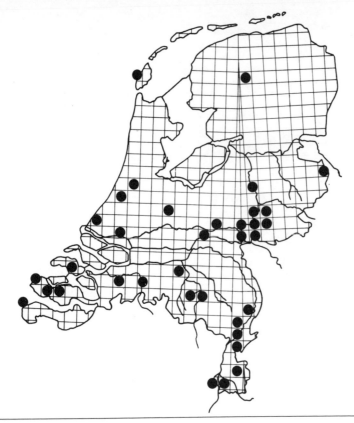

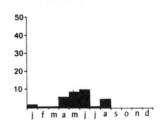

222 Bradycellus similis (Dej.)   *B. ruficollis Steph.*

macropterous   254 records

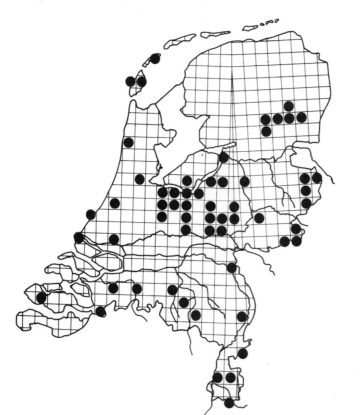

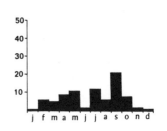

223 Bradycellus verbasci (Dft.)

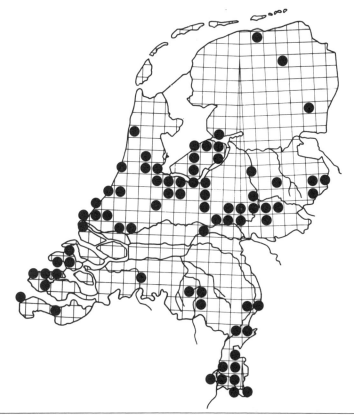

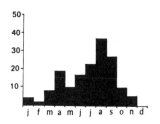

224 Bradycellus sharpi Joy

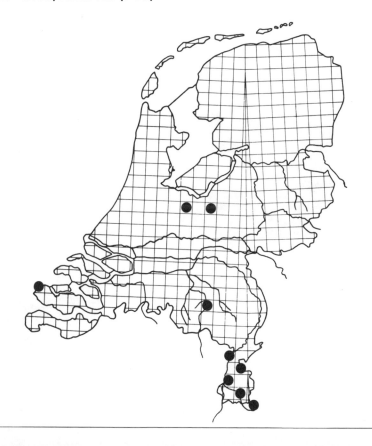

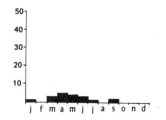

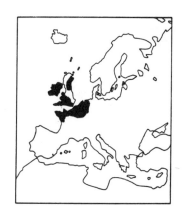

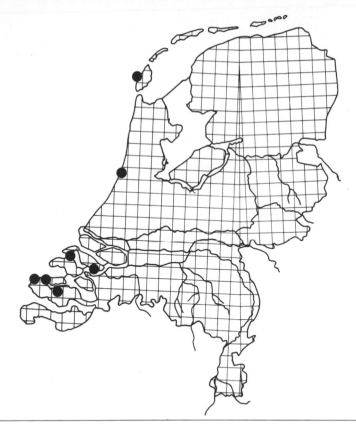

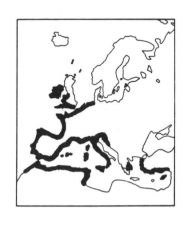

226 **Bradycellus harpalinus Serv.**   dimorphic   401 records

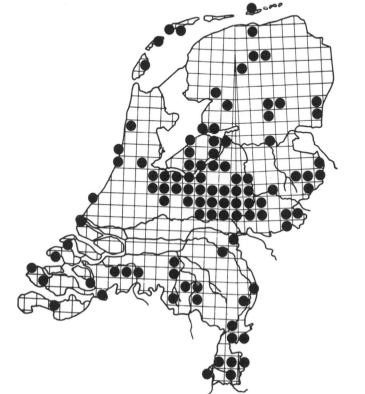

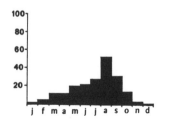

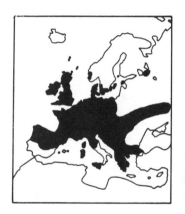

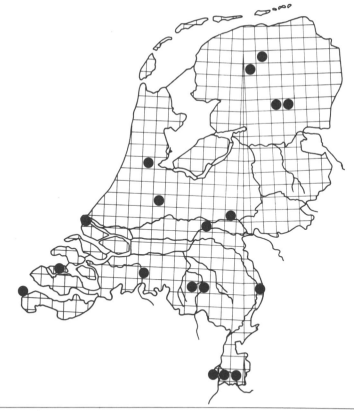

228   **Bradycellus collaris (Payk.)**   dimorphic   174 records

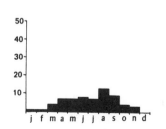

229 Trichocellus placidus (Gyll.)

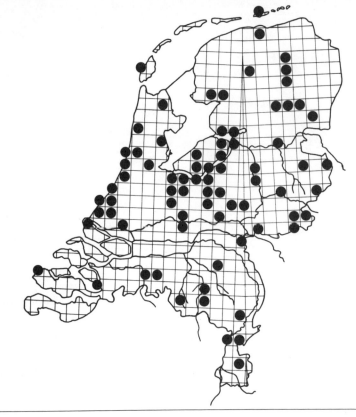

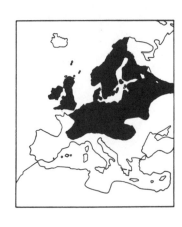

230 Trichocellus cognatus (Gyll.)

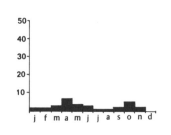

231   Dichirotrichus pubescens (Payk.)   *Dicheirotrichus gustavi Crotch*   macropterous   179 records

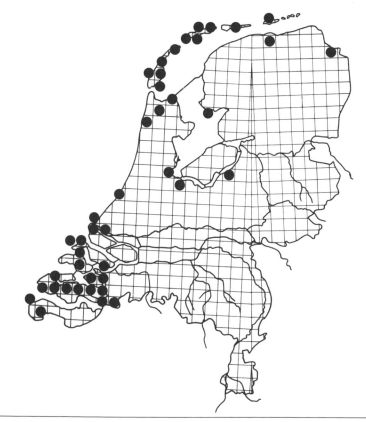

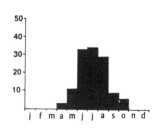

232   Dichirotrichus obsoletus (Dej.)   *Dicheirotrichus obsoletus (Dej.)*   macropterous   43 records

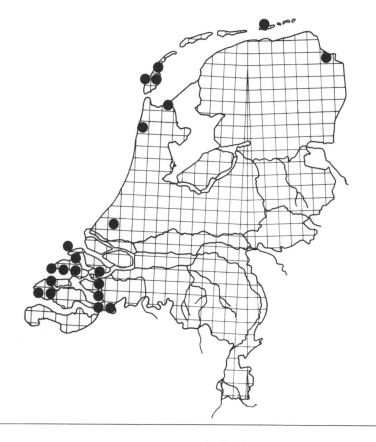

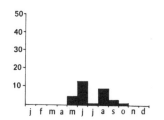

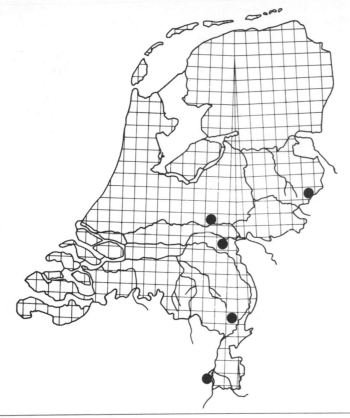

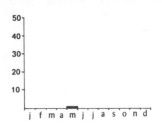

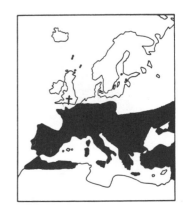

234   **Anisodactylus binotatus (F.)**   macropterous   460 records

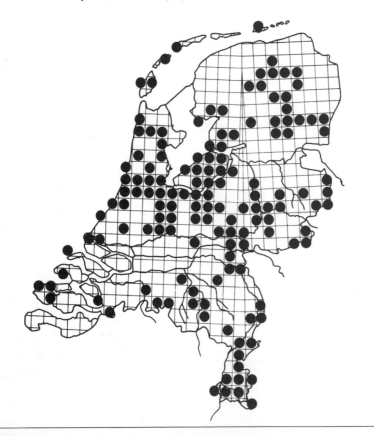

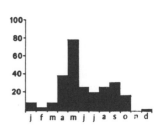

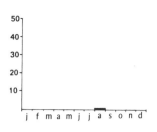

236   Anisodactylus signatus (Panz.)   13 records

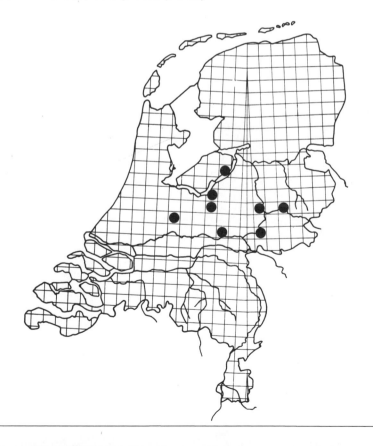

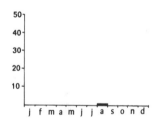

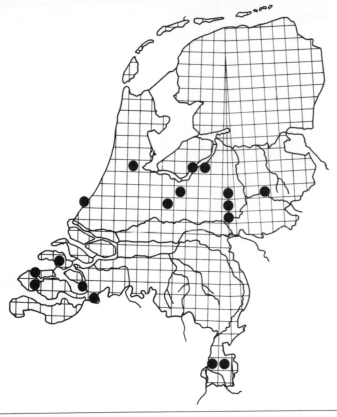

larvae

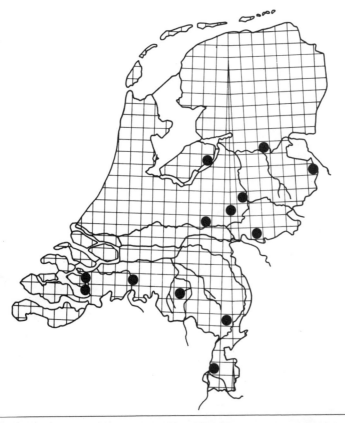

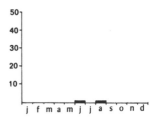

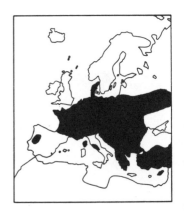

239 Amara plebeja (Gyll.)

macropterous   549 records

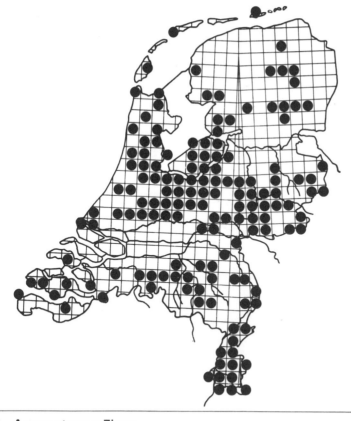

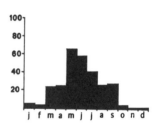

larvae

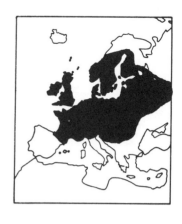

240 Amara strenua Zimm.

5 records

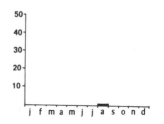

macropterous   6 records

242 **Amara similata (Gyll.)**

macropterous   289 records

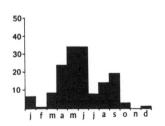

larvae

243 Amara ovata (F.)   macropterous 136 records

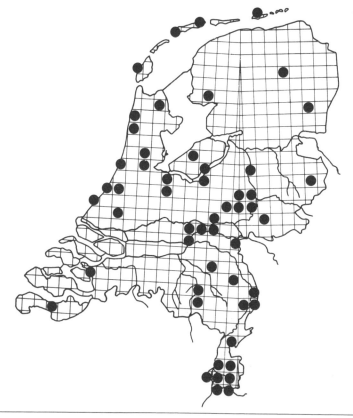
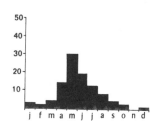

244 Amara montivaga Strm.   64 records

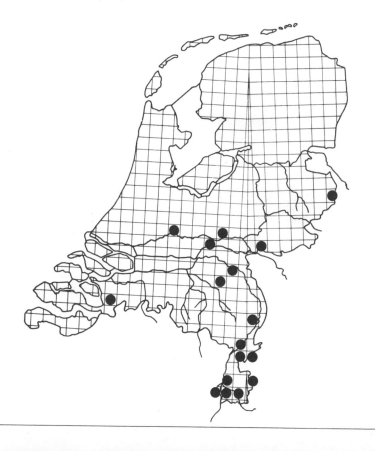
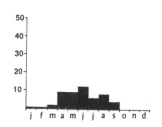

macropterous 8 records

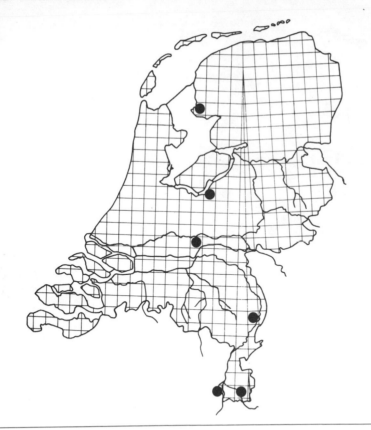

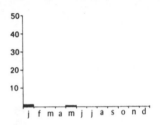

246 **Amara communis (Panz.)**

macropterous 569 records

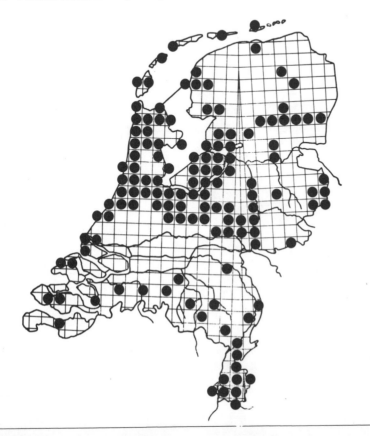

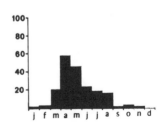

larvae

247 Amara convexior Steph.

macropterous   94 records

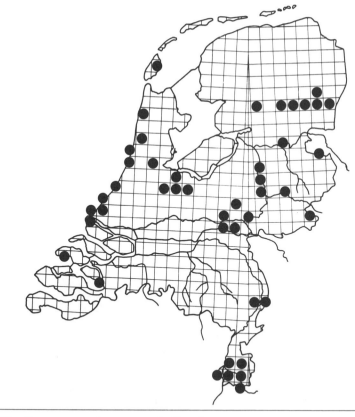

248 Amara pseudocommunis Burak

macropterous   38 records

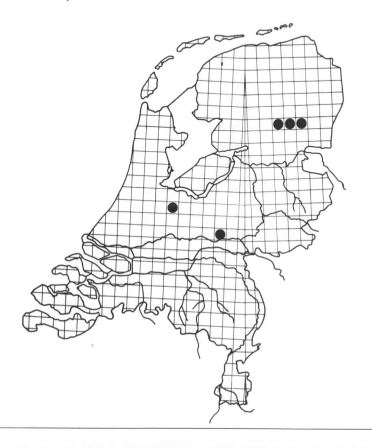

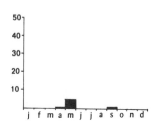

249 Amara lunicollis Schdte. *A. vulgaris (L.)*   macropterous 421 records

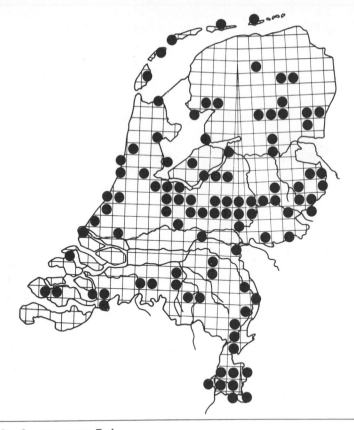

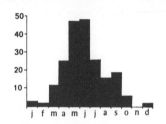

larvae

250 Amara curta Dej.   macropterous 155 records

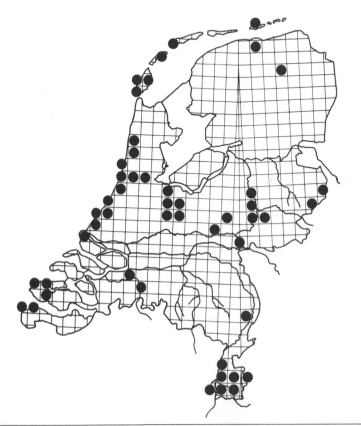

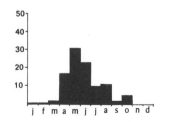

251 Amara aenea (Deg.)

macropterous   580 records

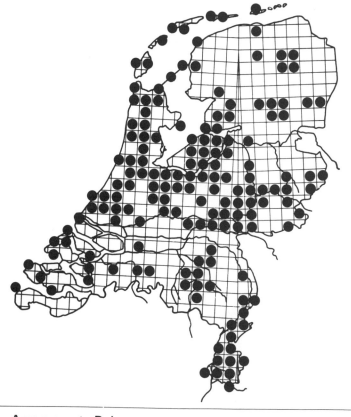

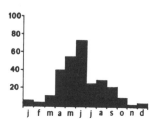

larvae

252 Amara spreta Dej.

macropterous   389 records

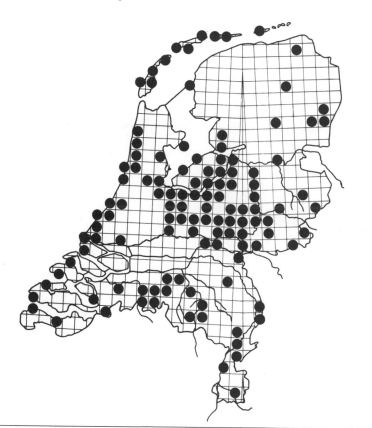

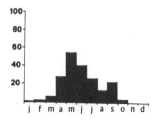

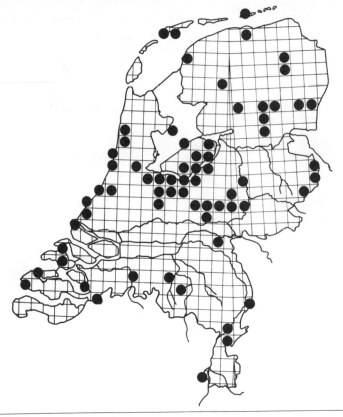

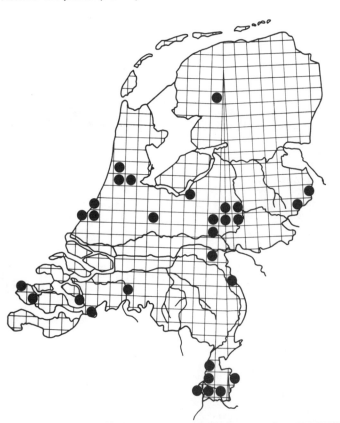

larvae

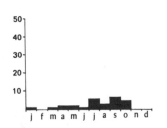

macropterous   457 records

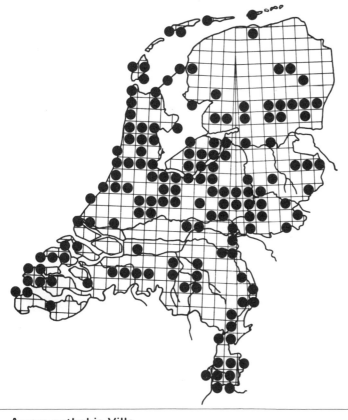

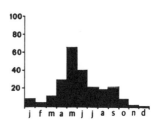

larvae

99 records

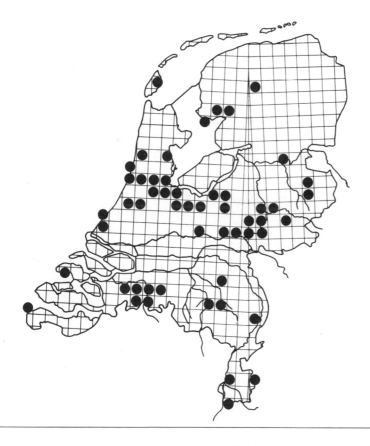

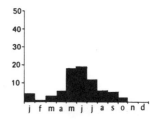

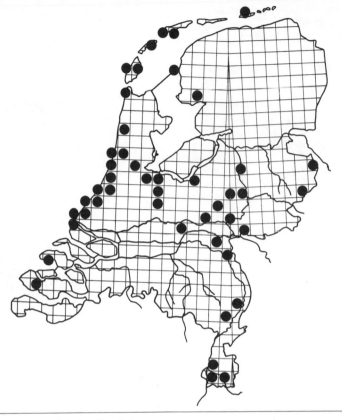

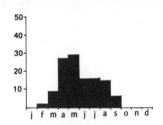

258 Amara tibialis (Payk.)   macropterous 110 records

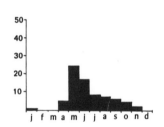

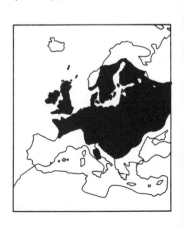

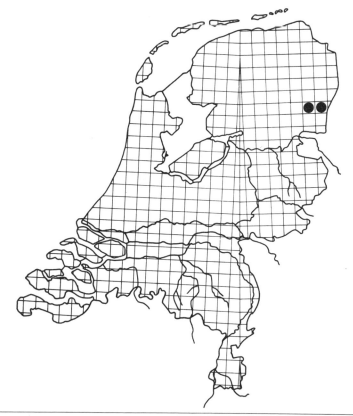

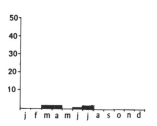

260 **Amara fusca Dej.**   macropterous   5 records

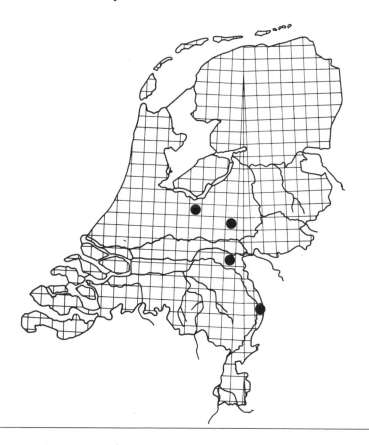

261 Amara cursitans Zimm.   macropterous   18 records

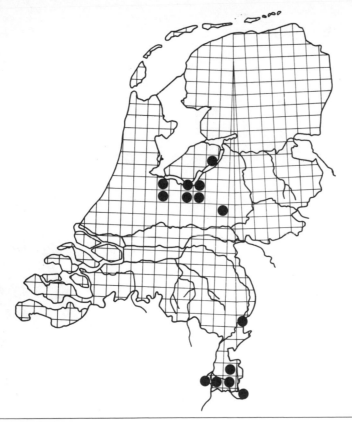

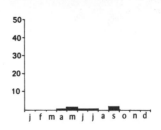

262 Amara silvicola Zimm. *A. quenseli (Schönh.)*   68 records

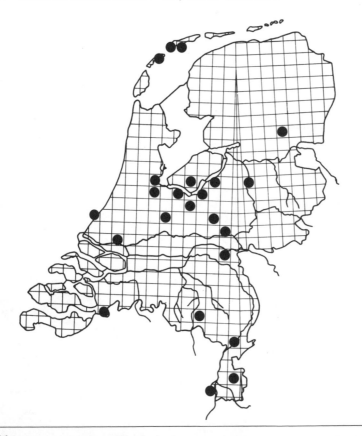

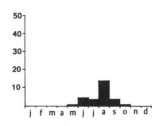

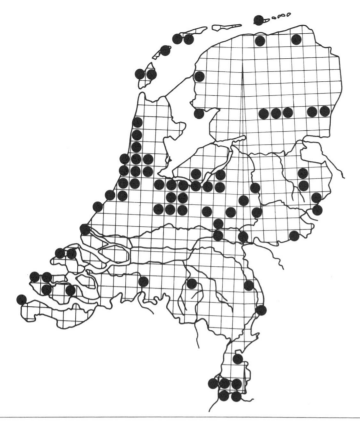

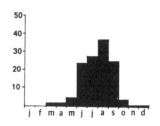

264 Amara infima (Dft.)   dimorphic   114 records

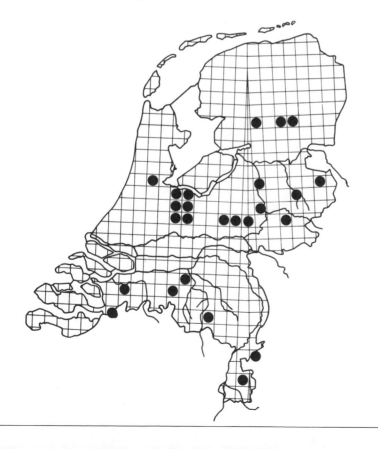

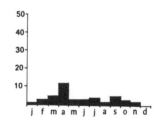

macropterous   66 records

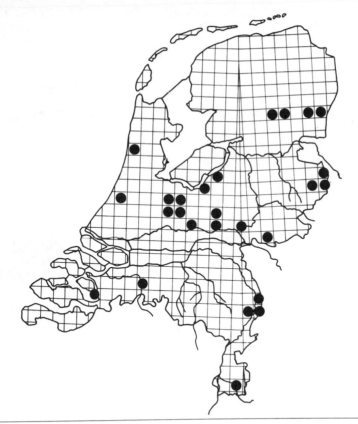

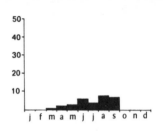

266   Amara brunnea (Gyll.)

macropterous   86 records

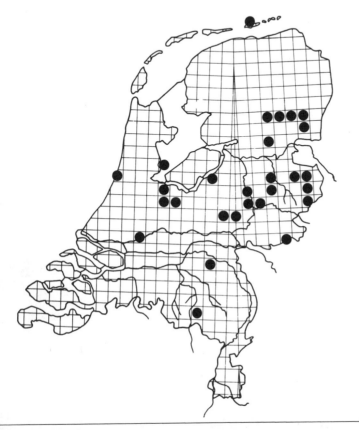

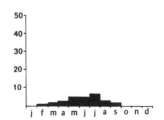

larvae

267 Amara apricaria (Payk.)

macropterous   198 records

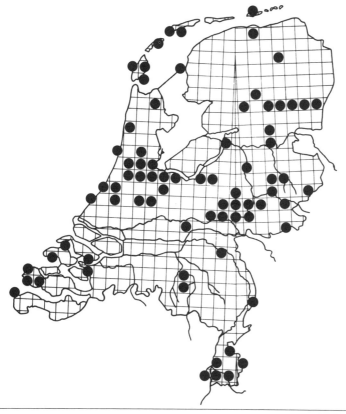

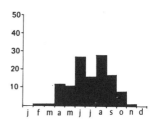

larvae

268 Amara fulva (Deg.)

macropterous   274 records

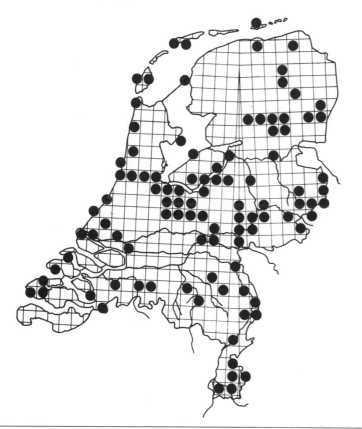

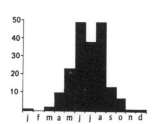

larvae

270   Amara consularis Dft.   macropterous   133 records

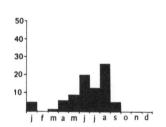

larvae

Amara aulica (Panz.)   macropterous   54 records

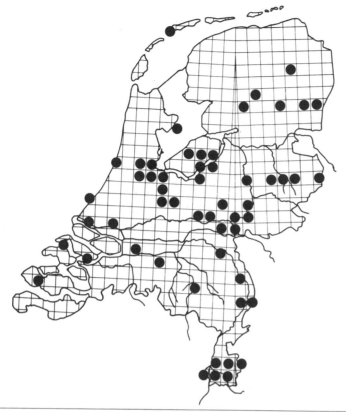
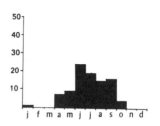

---

272 **Amara convexiuscula (Mrsh.)**   macropterous   167 records

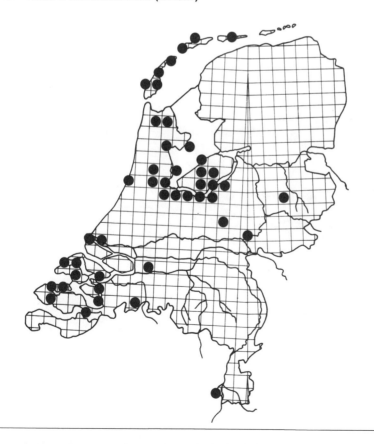
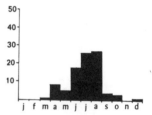

macropterous   121 records

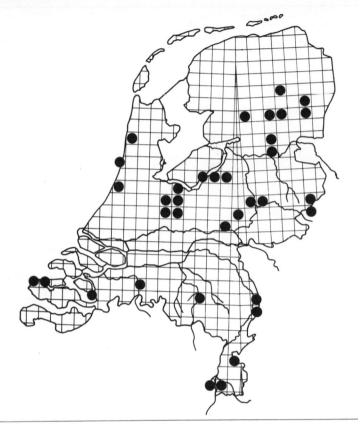

274   **Zabrus tenebrioides Goeze**

macropterous   91 records

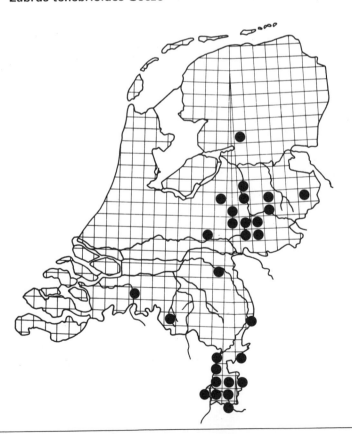

larvae

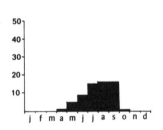

275 Stomis pumicatus (Panz.) dimorphic 259 records

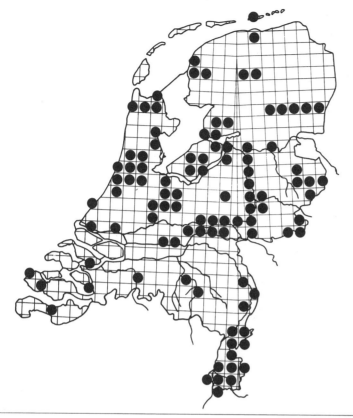

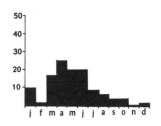

276 Pterostichus punctulatus (Schall.) macropterous 6 records

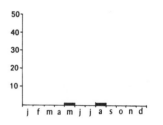

171

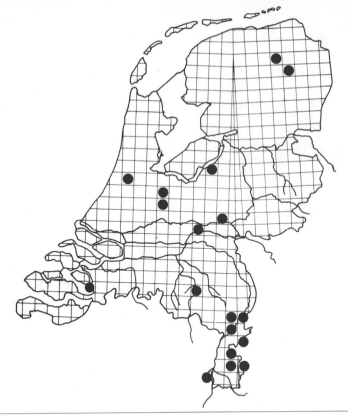
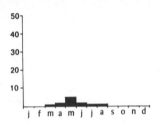

278   Pterostichus lepidus (Leske)   dimorphic   439 records

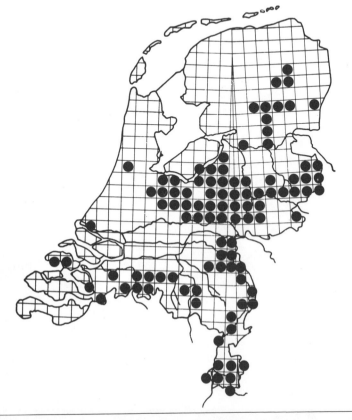
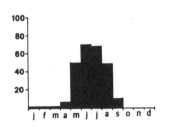

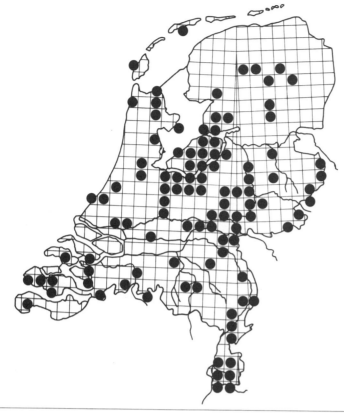

larvae

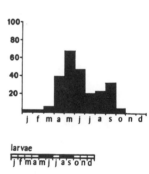

280   **Pterostichus coerulescens (L.)**   *P. versicolor Sturm.*   macropterous   566 records

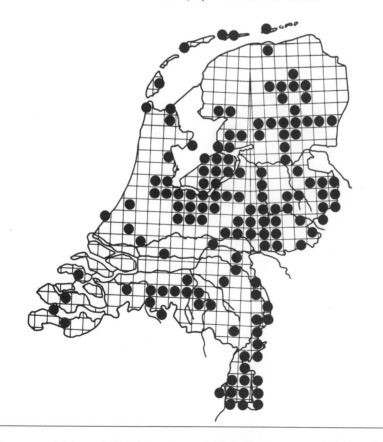

larvae

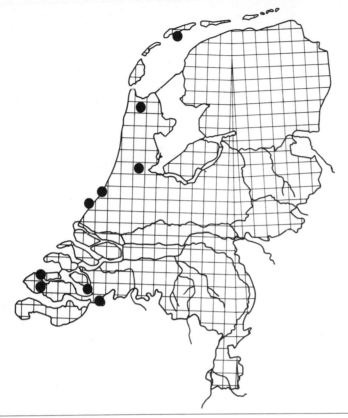

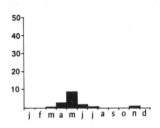

282 **Pterostichus vernalis (Panz.)**   dimorphic   577 records

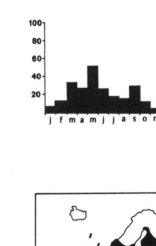

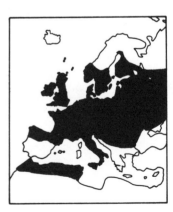

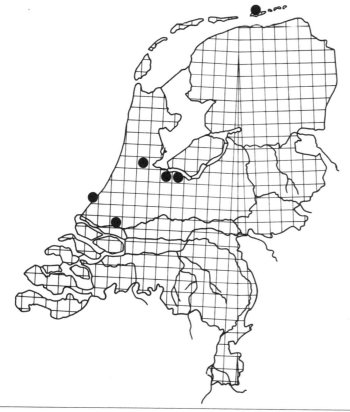

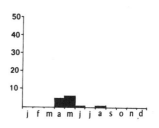

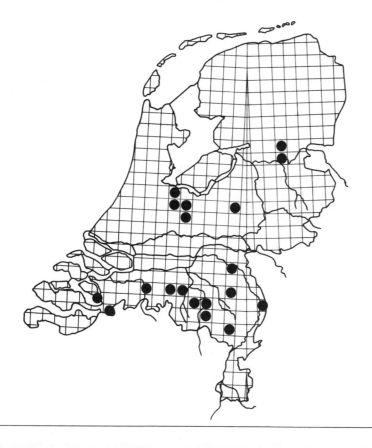

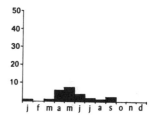

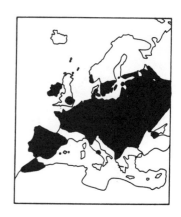

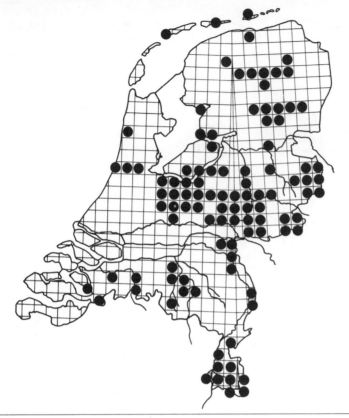

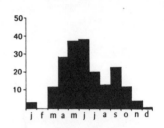

larvae

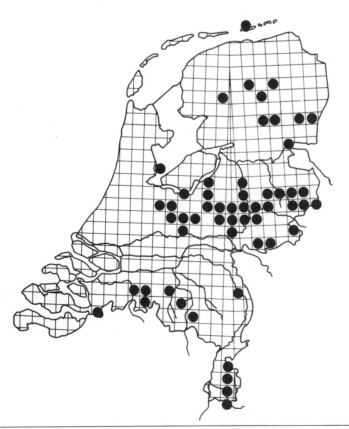

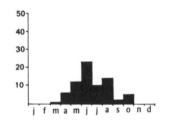

Pterostichus niger (Schall.)   macropterous   615 records

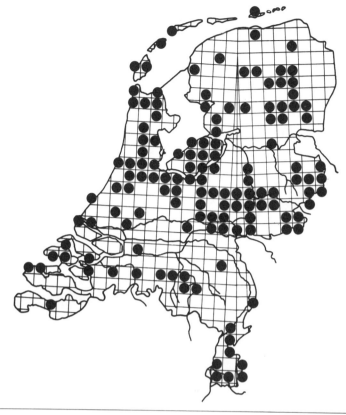

larvae

288   **Pterostichus vulgaris L.**   *P. melanarius Ill.*   dimorphic   827 records

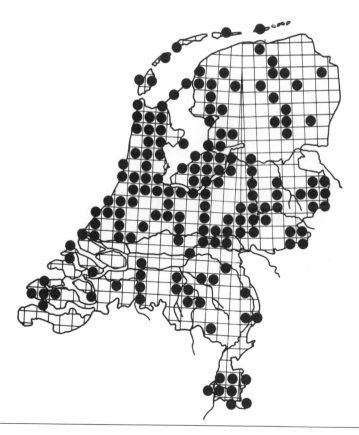

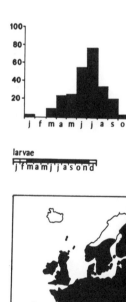

larvae

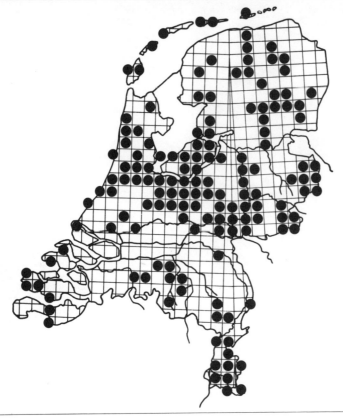

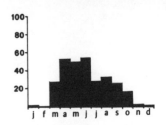

290   Pterostichus anthracinus (III.)   dimorphic   142 records

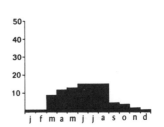

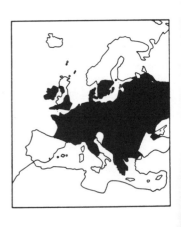

291 Pterostichus gracilis Dej.

macropterous 37 records

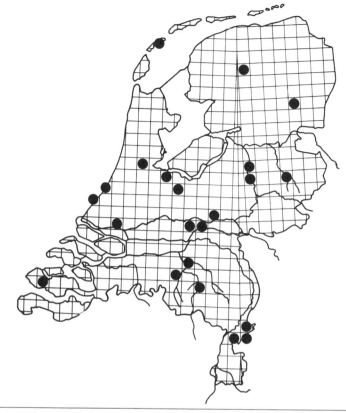

---

292 Pterostichus minor (Gyll.)

dimorphic 325 records

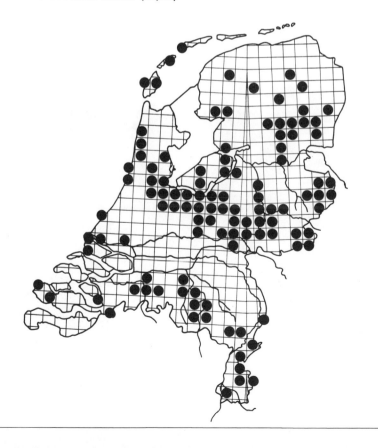

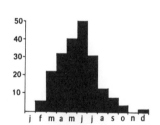

larvae

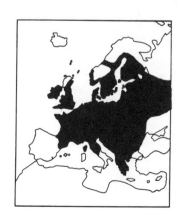

dimorphic   670 records

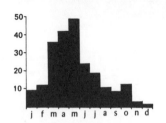

larvae

294   Pterostichus diligens Strm.

dimorphic   496 records

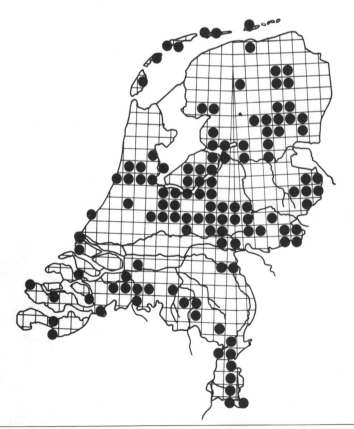

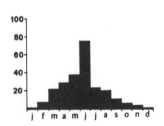

larvae

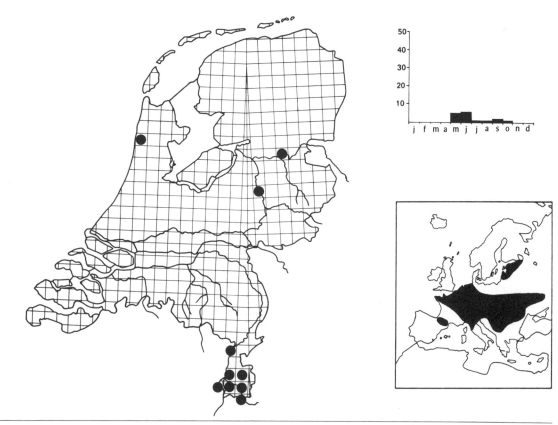

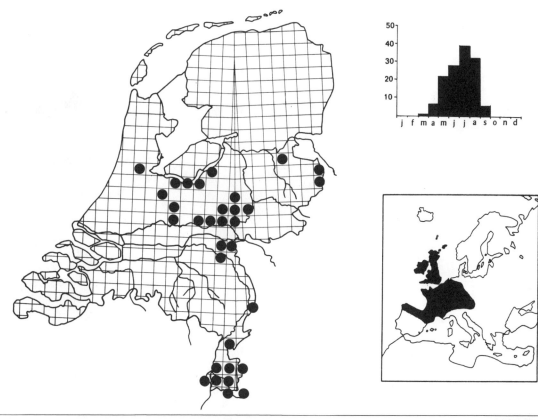

297 Pterostichus cristatus (Duf.)

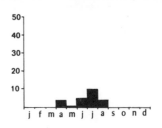

298 Abax ater (Villers) *A. Parallelepipedus (Pill. et Mitp.)*

brachypterous   252 records

larvae

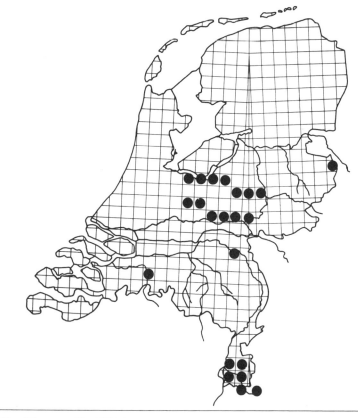
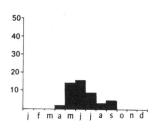
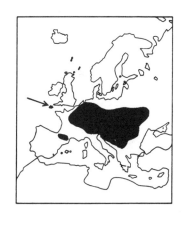

300 **Abax ovalis (Dft.)**   30 records

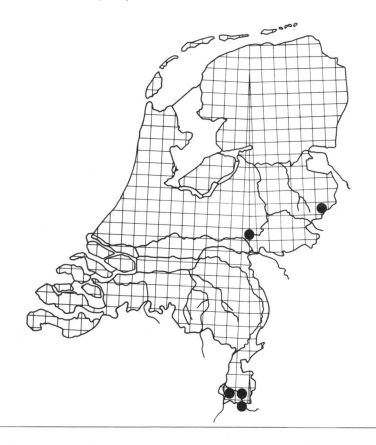
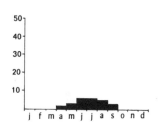

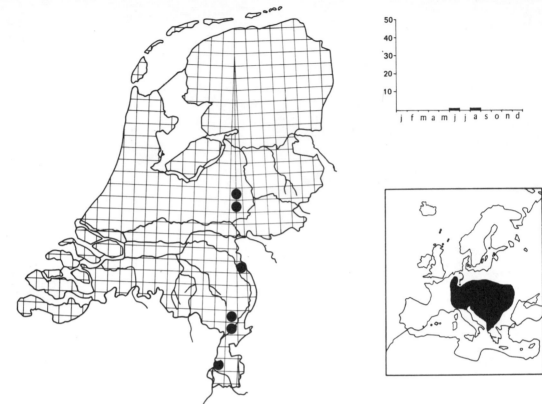

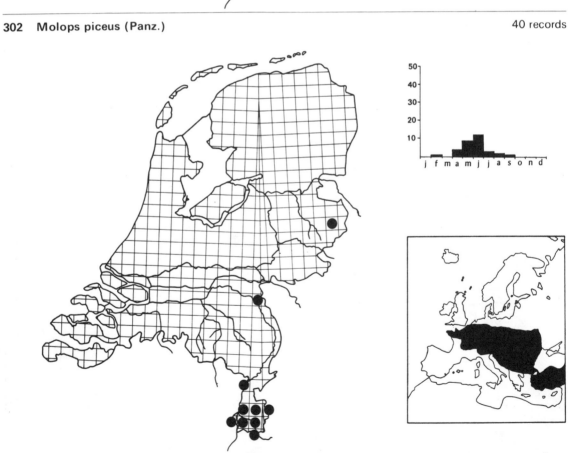

303 Calathus fuscipes (Goeze)

brachypterous   389 records

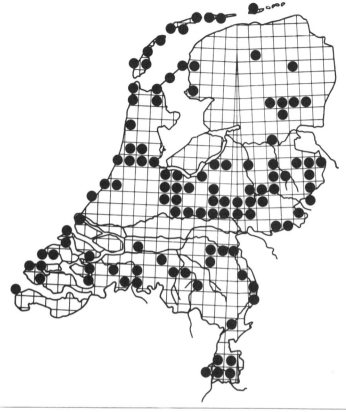

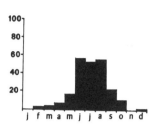

larvae

304 Calathus erratus Sahlb.

dimorphic   461 records

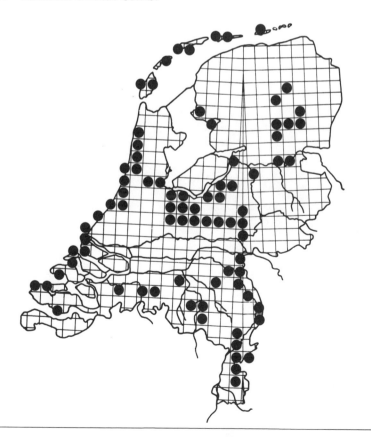

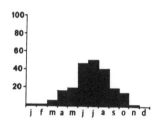

larvae

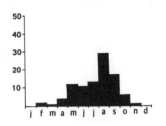

306 Calathus micropterus (Dft.) brachypterous 150 records

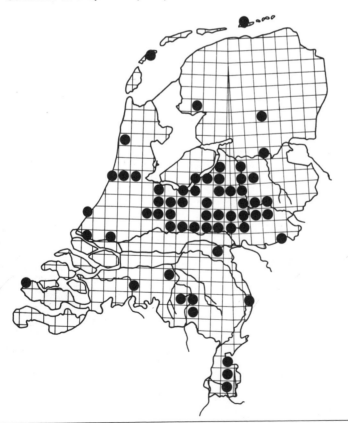

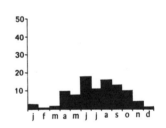

larvae

307   Calathus melanocephalus (L.)

dimorphic   616 records

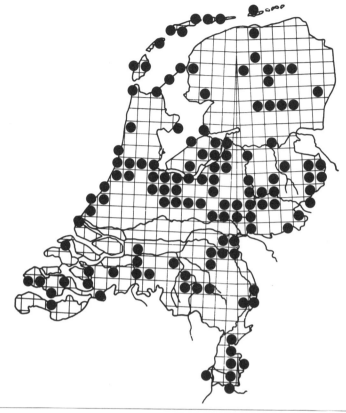

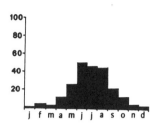

larvae

308   Calathus mollis (Mrsh.)

dimorphic   176 records

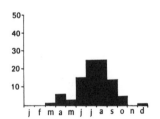

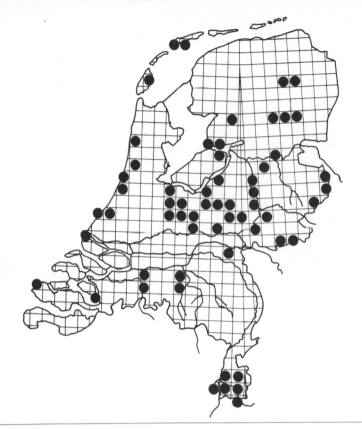

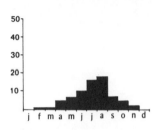

310 **Sphodrus leucophthalmus (L.)**   macropterous   6 records

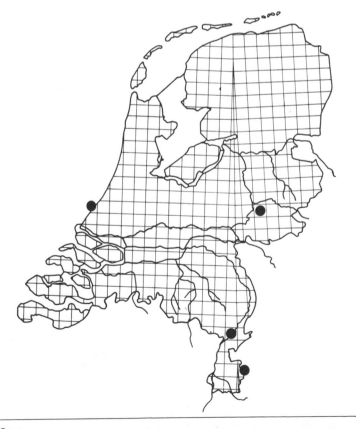

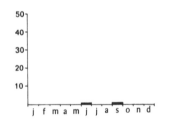

311 Pristonychus terricola (Hbst.)

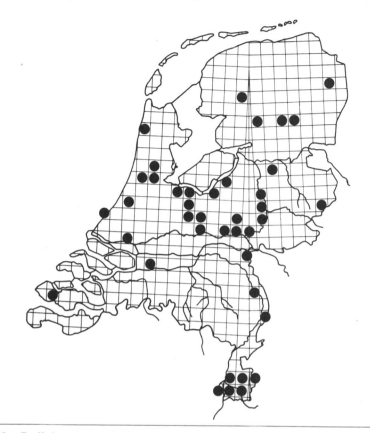

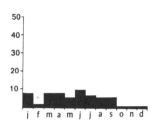

312 Dolichus halensis (Bon.)

macropterous   4 records

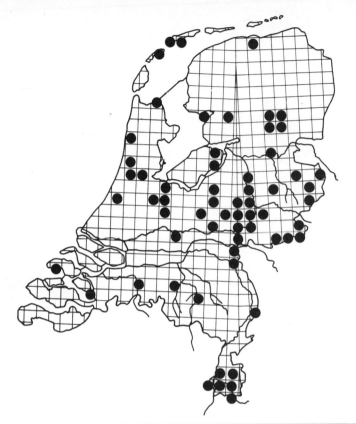

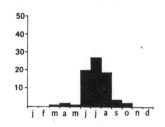

314 Olistopus rotundatus (Payk)   dimorphic   138 records

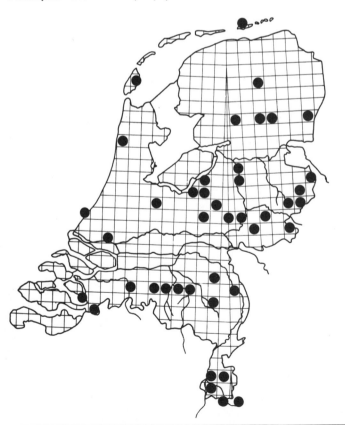

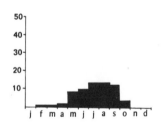

larvae

315   Agonum impressum (Panz.)   2 records

316   Agonum sexpunctatum (L.)   macropterous   332 records

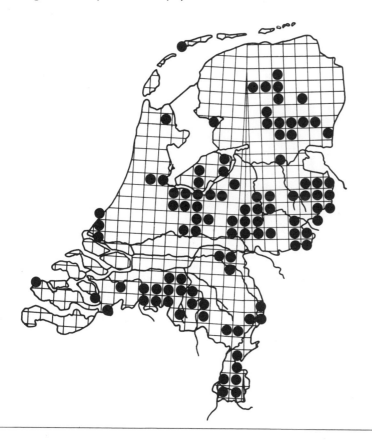

larvae

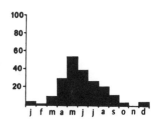

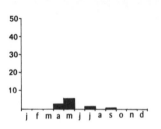

318 Agonum viridicupreum (Goeze)   18 records

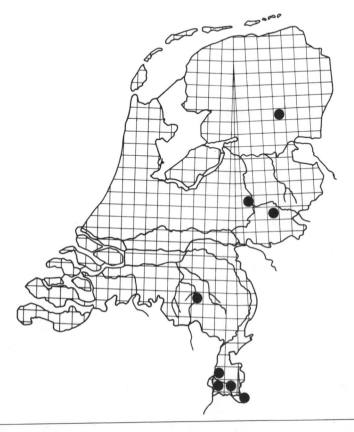

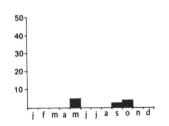

319 Agonum gracilipes (Dft.)

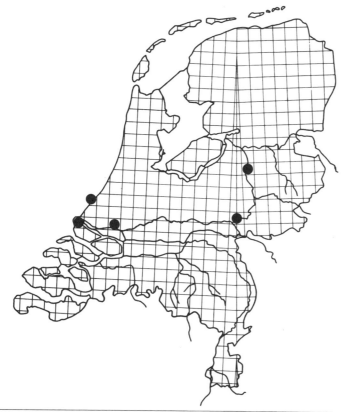

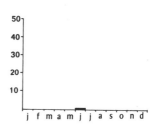

320 Agonum marginatum (L.)

macropterous   291 records

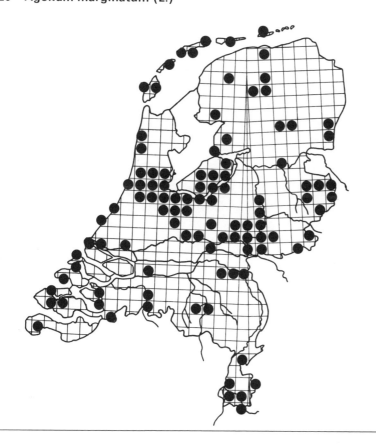

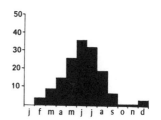

larvae

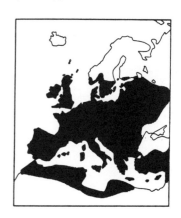

321 Agonum muelleri (Hbst.)

macropterous 339 records

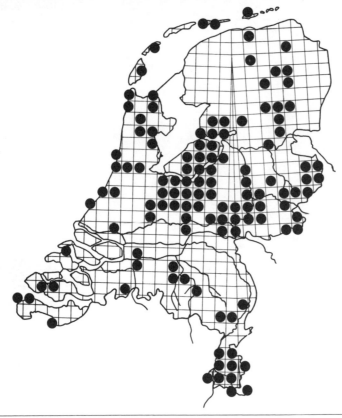

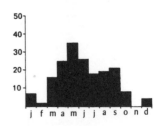

larvae

322 Agonum dolens (Sahlb.)

macropterous 24 records

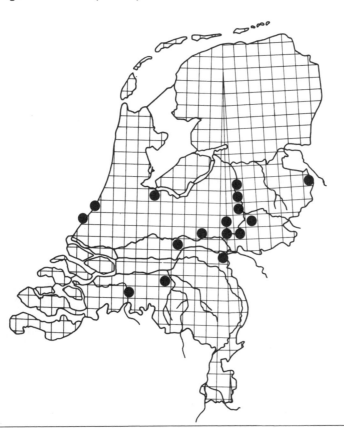

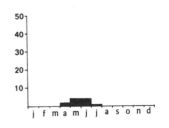

323 Agonum versutum (Gyll.)

macropterous 97 records

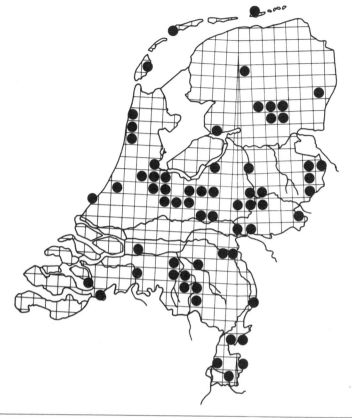

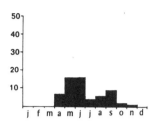

324 Agonum viduum (Panz.)

macropterous 149 records

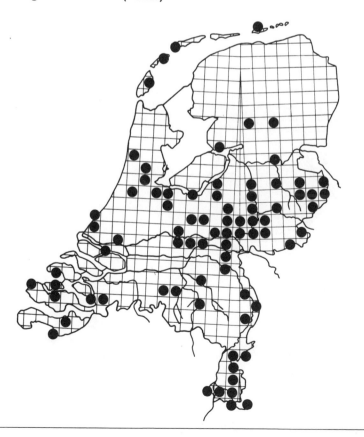

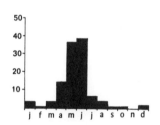

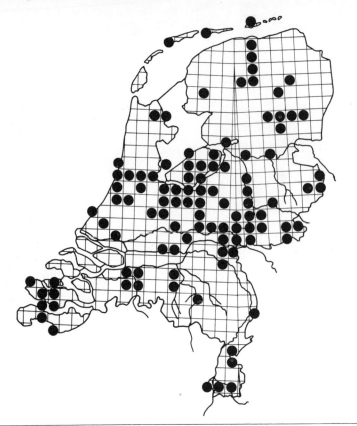

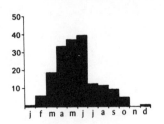

326   Agonum dahli Prdh.   *A. nigrum Dej.*   8 records

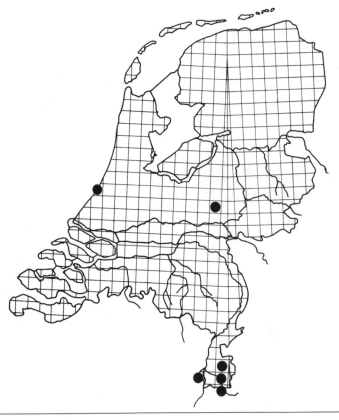

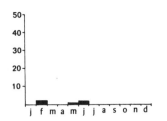

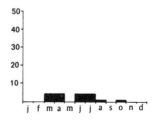

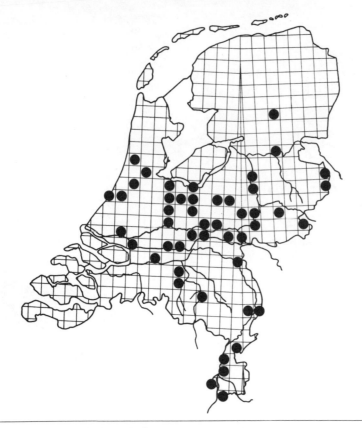

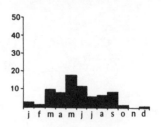

330 **Agonum scitulum Dej.**   21 records

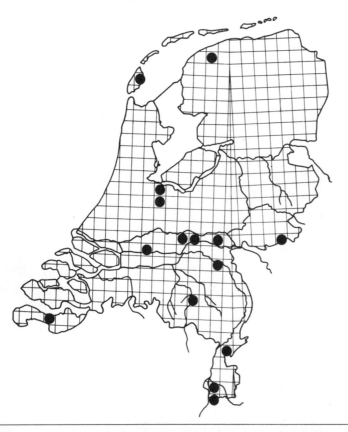

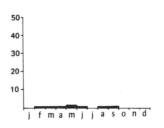

331 Agonum piceum (L.)

macropterous   65 records

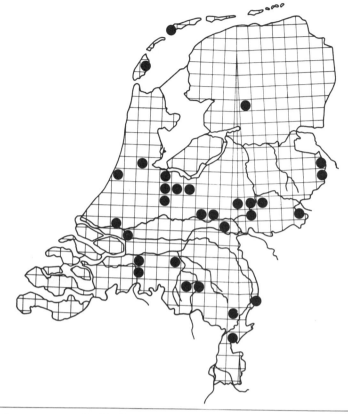

332 Agonum gracile (Gyll.)

macropterous   89 records

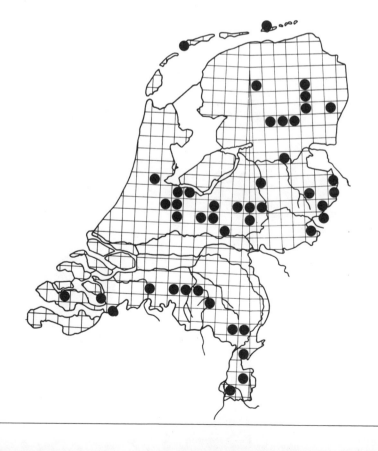

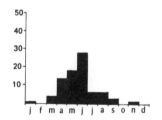

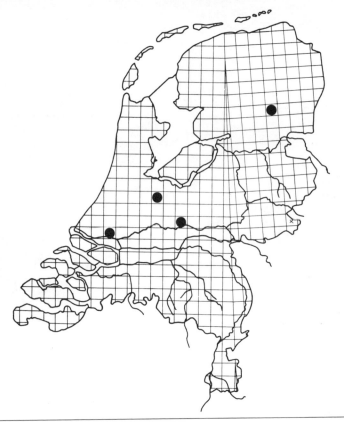

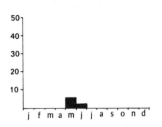

334 **Agonum fuliginosum (Panz.)**

dimorphic   269 records

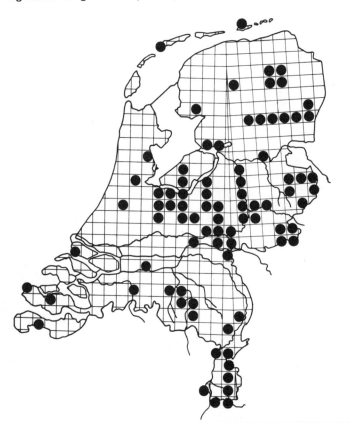

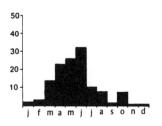

larvae

335 Agonum thoreyi Dej. *A. puellum Dej.*

macropterous   192 records

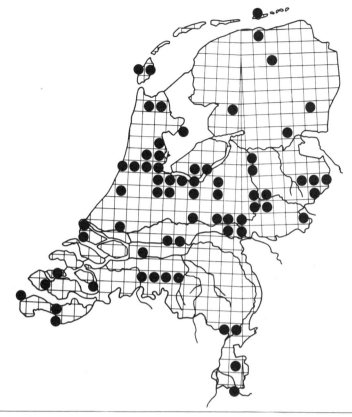

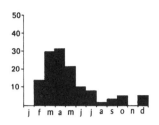

336 Agonum assimile (Payk.)

macropterous   303 records

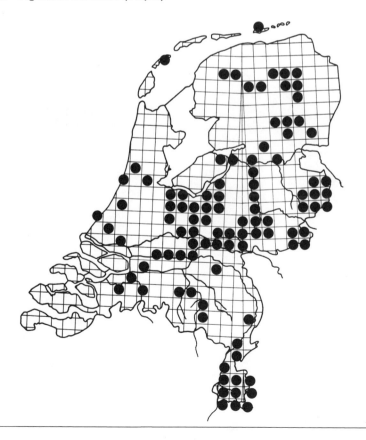

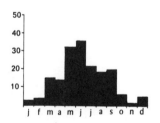

larvae

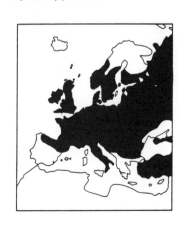

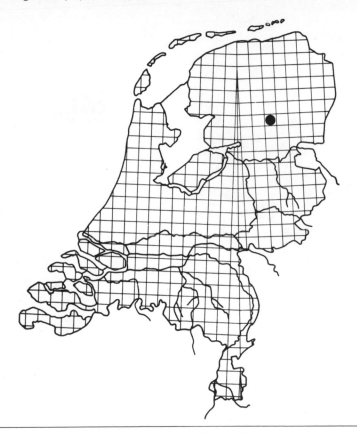

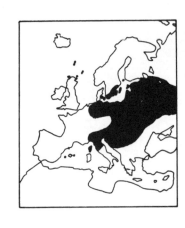

338 **Agonum ruficorne Goeze** *A. albipes (F.)* macropterous 220 records

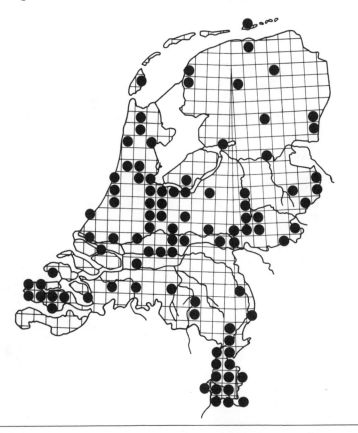

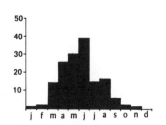

larvae

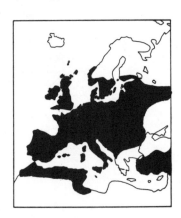

339   Agonum obscurum Hbst.   dimorphic   429 records

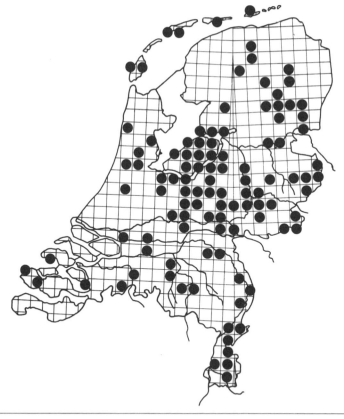

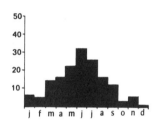

larvae

340   Agonum dorsale (Pont.)   macropterous   437 records

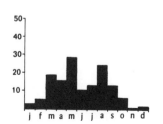

larvae

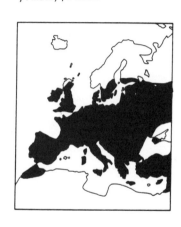

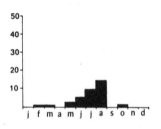

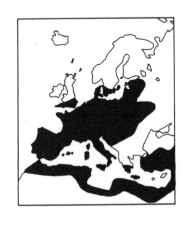

342   Lebia chlorocephala (Hoffm.)   macropterous   99 records

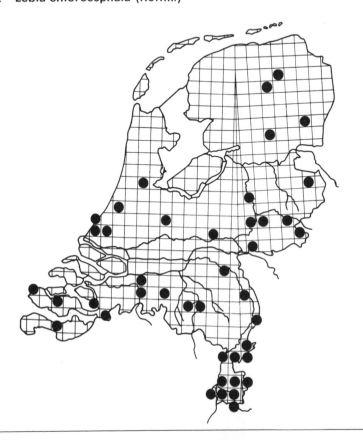

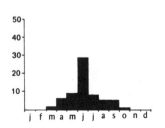

larvae

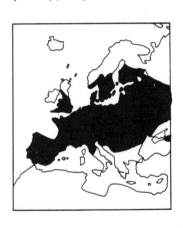

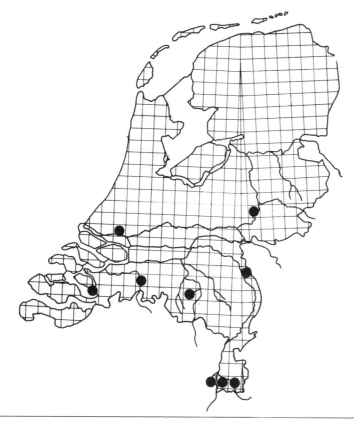

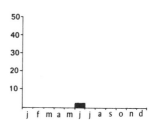

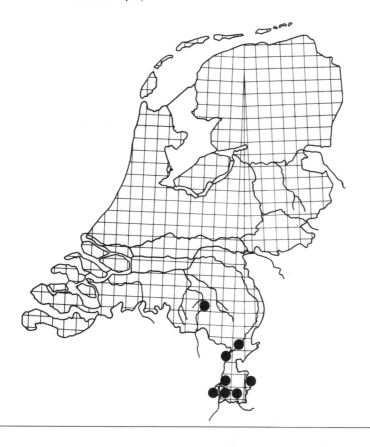

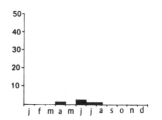

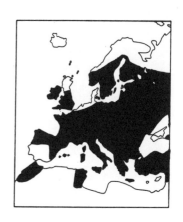

345 Demetrias atricapillus (L.)

macropterous   194 records

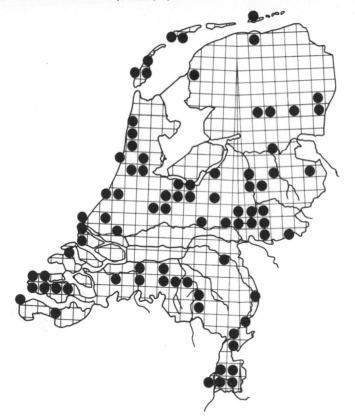

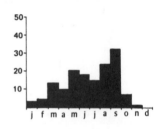

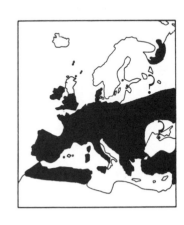

346 Demetrias monostigma Sam.

brachypterous   152 records

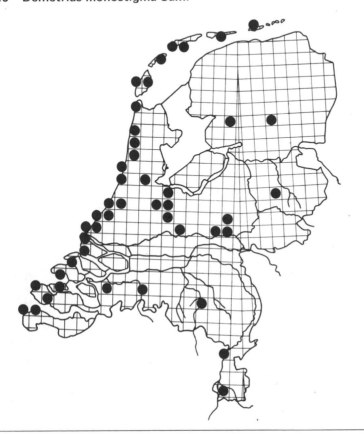

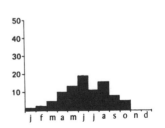

larvae

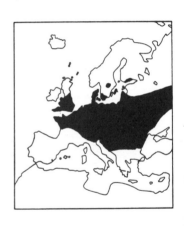

347 Demetris imperialis (Germ.)

macropterous   71 records

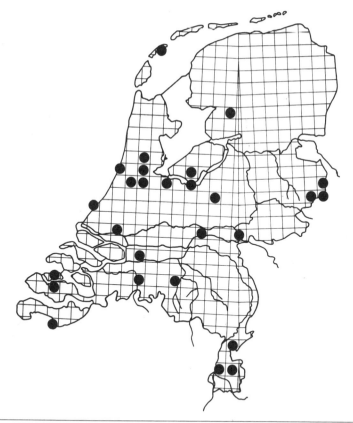

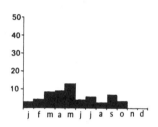

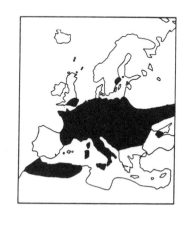

348 Dromius longiceps Dej.

macropterous   13 records

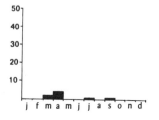

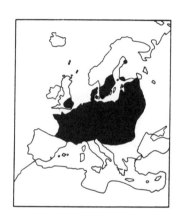

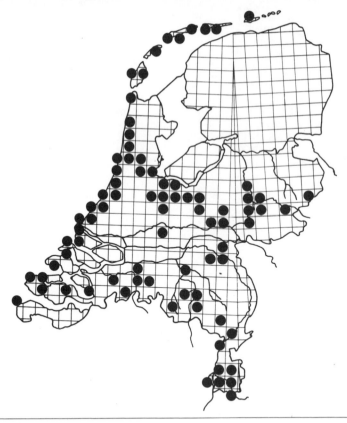

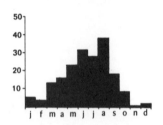

larvae

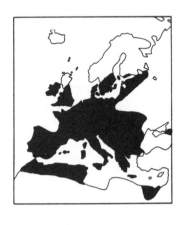

350 Dromius agilis (F.)   macropterous   141 records

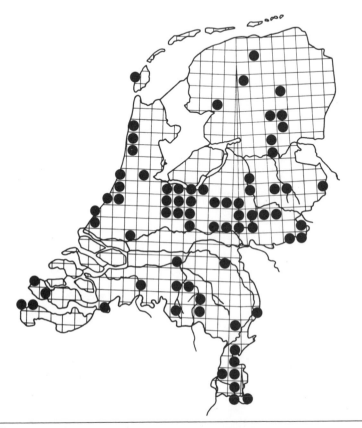

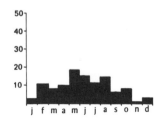

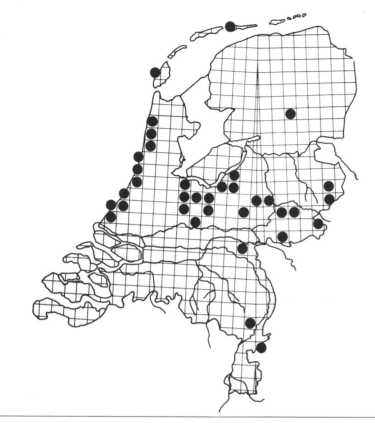

352 Dromius meridionalis Dej.   9 records

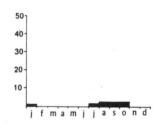

352.1 Dromius marginellus (F.)

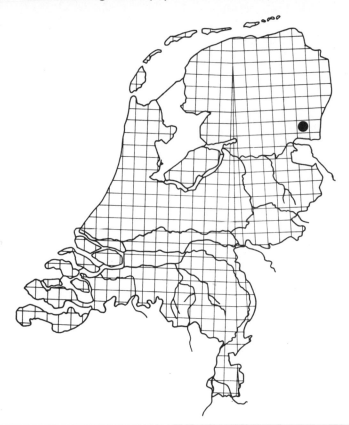

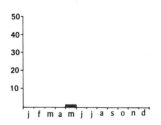

353 Dromius fenestratus (F.)

macropterous   4 records

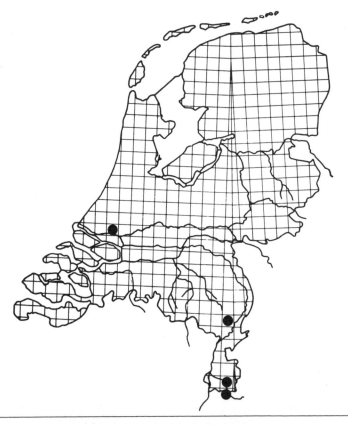

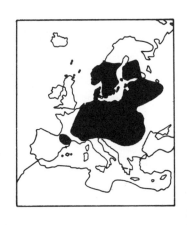

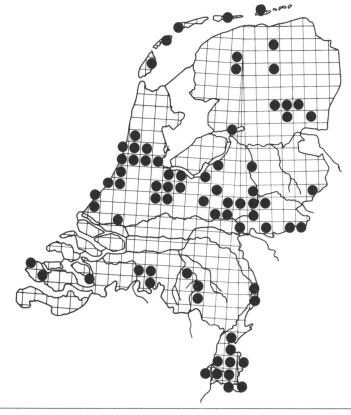

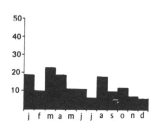

larvae

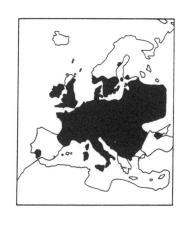

355   Dromius quadrinotatus (Panz.)   macropterous   230 records

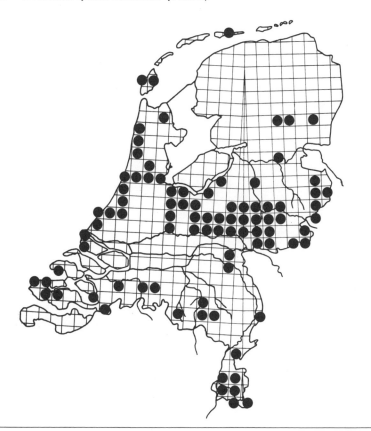

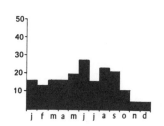

356 Dromius quadrisignatus Dej.

5 records

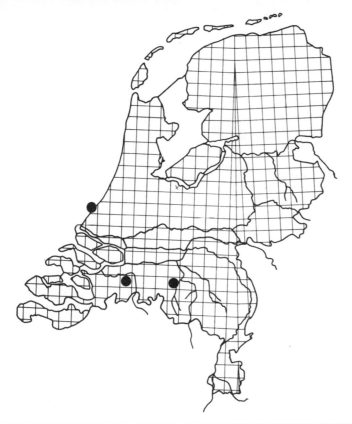

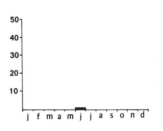

357 Dromius sigma (Rossi)

dimorphic   62 records

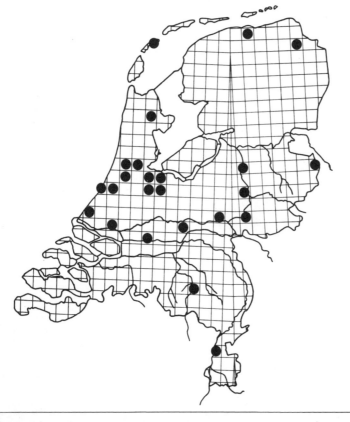

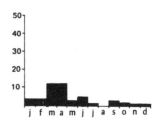

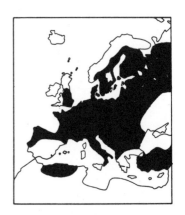

Dromius nigriventris Dej.   *D. notatus Steph.*   dimorphic   48 records

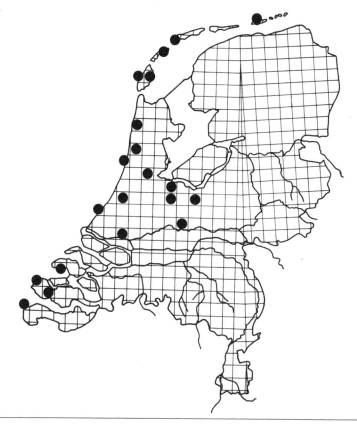

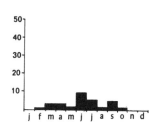

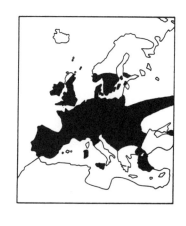

359 **Dromius melanocephalus Dej.**   macropterous   185 records

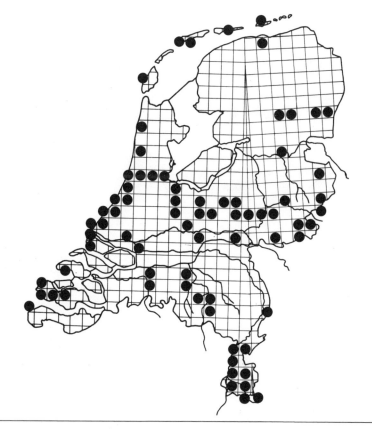

361   **Metabletus truncatellus (L.)**   dimorphic   159 records

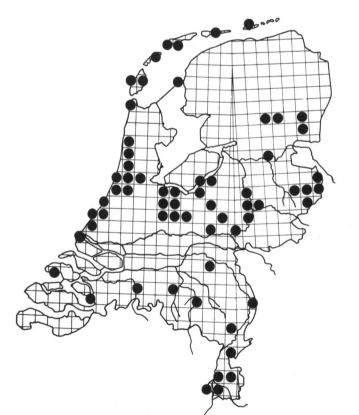

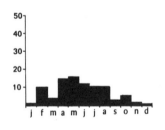

362   Metabletus foveatus (Fourcr.)

brachypterous   421 records

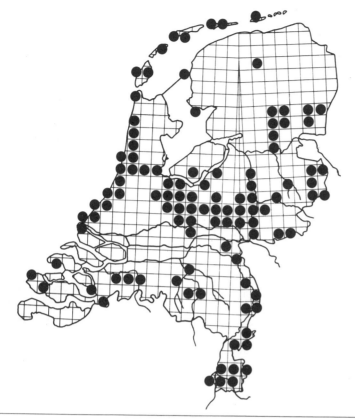

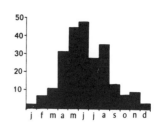

363   Microlestes minutulus (Goeze)

macropterous   60 records

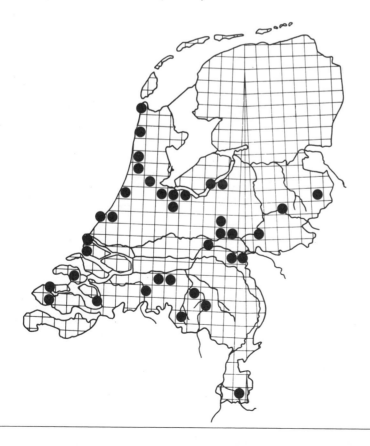

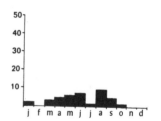

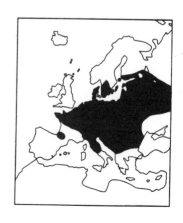

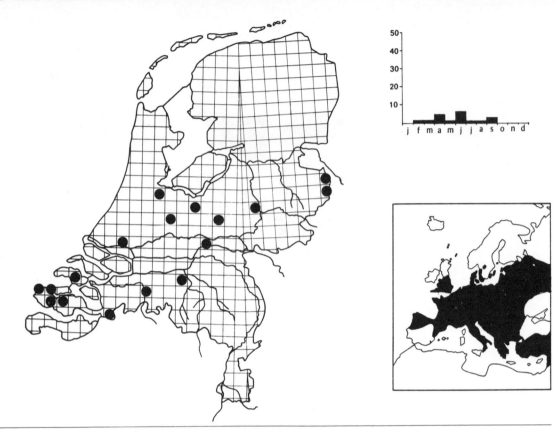

365 **Lionychus quadrillum (Dft.)**   15 records

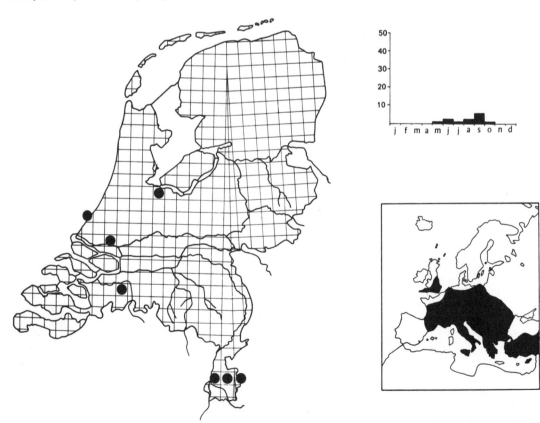

366 Cymindis humeralis (Fourcr.)

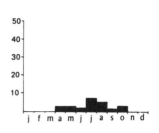

367 Cymindis axillaris F.

14 records

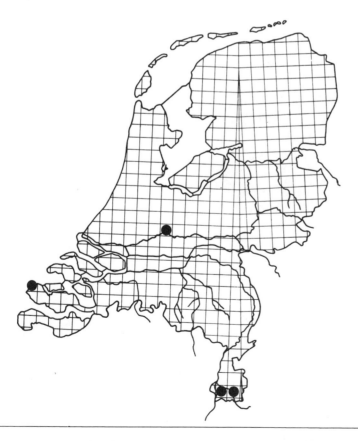

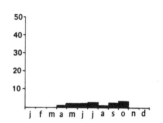

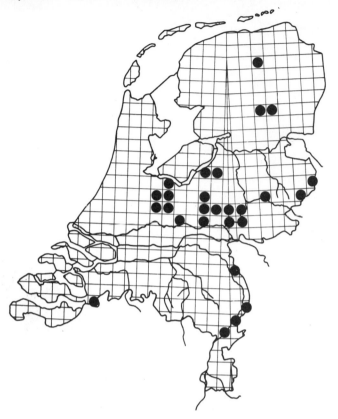

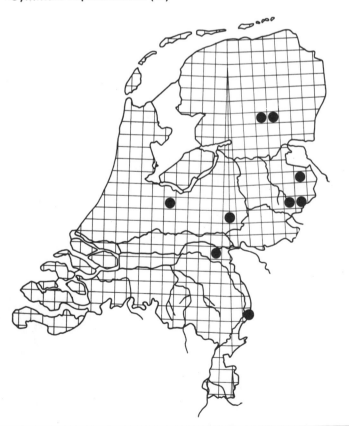

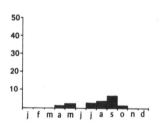

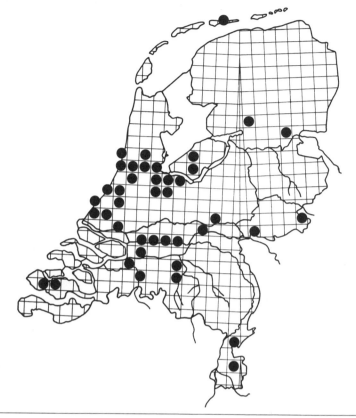

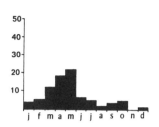

larvae

---

371   Brachynus crepitans L.   56 records

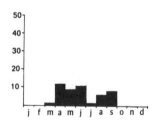

larvae

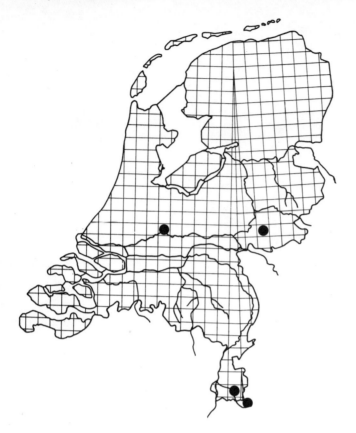

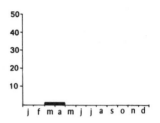

Appendices

A. Notes on the computer programme

The programme, which is in PL/I, makes it possible to produce, for any set of data and for any given area or country, a printed map showing the frequencies with which a particular species has been recorded (see Figure 2).

We used three types of cards, one of which (type 1) carries the individual data, the others being reserved for information of a general nature. *Card type 1* bears information derived from collections, the literature, or the field. The inclusion of part of this information is obligatory, the rest being optional. Insertion of the full name of the sampling site provides a useful check on punching errors concerning the row and column number of the grid. Information about the source is also helpful in the checking of doubtful records.

Card type 2 relates the species number on card type 1 to the species name and, if desired, to biological information on the species in question.

Card type 3 defines the area covered in the mapping investigation. Since we used a printer instead of a plotter, the output is in matrix form, and because the area covered is generally not rectangular, a matrix element outside the area under study must be suppressed. The total number of columns in the matrix should not exceed 33.

The IN data set comprises:

On card type 1

columns	
1– 3	species number
4– 5	row number of the grid
6– 7	column number of the grid
8–11	internal (alpha-numeric) indication of the grid (optional)
12–26	name of the locality (optional)
27	decade –0: up to 1900, 1: 1900–1910, ... 8: 1970–1980, 9: unknown.
28–29	month
30–31	source (optional)

The INHF8 data set comprises:

On card type 3

Matrix constituted row by row (one card for each row)

0	in a location outside the area under consideration (with no possibility to record an item)
1	in a location inside this area

The INHF3 data set comprises:

On card type 2

columns	
1– 3	species number
4– 6	species number for internal use if different from the first one (e.g., number based on alphabetical order instead of the systematic one) (optional)
7–40	codes of biological information (optional)
41–60	genus name
61–80	species name

B. Computer programme

```
KRT1: PROC OPTIONS(MAIN);

DCL

1 F BASED(S),
2 SNB CHAR(3),
2 CNR CHAR(2),
2 CNK CHAR(2),
2 GNR CHAR(4),
2 PLN CHAR(15),
2 DTY CHAR(1),
2 DTM CHAR(2),
2 BRN CHAR(2);

1 HF3 BASED(P),
2 SNB CHAR(3),
2 SNP CHAR(3),
2 RST CHAR(34),
2 GEN CHAR(20),
2 SON CHAR(20);

SNP(374) CHAR(3),
GEN(374) CHAR(20),
SON(374) CHAR(20);

T(32,28) FIXED BIN,
TH(32,28) FIXED BIN,
(TM1(12),TM2(12)) FIXED BIN,
(DE,DL) FIXED BIN,
(ISNB,ISNV,ISW,I,J,K) FIXED BIN;

SNP,GEN,SON=' ';

DO I=1 TO 32; DO J=1 TO 28; GET FILE(INHF8) LIST(TH(I,J)); END; END;

ON ENDFILE(INHF3) GO TO VERDER1;

LABA;
READ FILE(INHF3) SET(P);

GET STRING(HF3.SNB) EDIT(ISNB)(F(3));

SNP(ISNB)=HF3.SNP;

GEN(ISNB)=HF3.GEN;
SON(ISNB)=HF3.SON;

GO TO LABA;

VERDER1;
T=0; TM1=0; TM2=0; ISW=0; ISNV=1; DE=8; DL=0;

ON ENDFILE(IN) BEGIN; ISW=1; GO TO PRINT; END;

LABB;
READ FILE(IN) SET(S);
GET STRING(F.SNB) EDIT(ISNB)(F(3));
IF ISNB¬=ISNV THEN GO TO PRINT;

LABC;
GET STRING(F.DTY) EDIT(I)(F(1));
IF I<DE THEN DE=I;
IF(I¬=9)&(I>DL) THEN DL=I;

GET STRING(F.DTM) EDIT(I)(F(2));
IF I¬=0 THEN DO;
GET STRING(F.BRN) EDIT(J)(F(2));
IF(J>7)&(J<14) THEN TM1(I)=TM1(I)+1;
ELSE TM2(I)=TM2(I)+1;
END;

GET STRING(F.CNR) EDIT(I)(F(2));
GET STRING(F.CNK) EDIT(J)(F(2));
T(I,J)=T(I,J)+1;

GO TO LABB;


PRINT:
PUT SKIP EDIT('  SPECIESNUMBER ACCORDING TO BRAKMAN : ',ISNV)
(A(40),F(3));
PUT SKIP EDIT('  GENUS AND SPECIESNAME : ',GEN(ISNV),SON(ISNV))
(A(40),A(20),A(20));
PUT SKIP(2) EDIT('  POOL  JAN.  FEBR  MRT.  APRL  MAY.  JUNE  JULY  AU
G.  SEPT  OKT.  NOV.  DEC.')(A);
PUT SKIP EDIT('   1  ')(A(6));
DO I=1 TO 12; PUT EDIT(TM1(I))(F(6)); END;
PUT SKIP EDIT('   2  ')(A(6));
DO I=1 TO 12; PUT EDIT(TM2(I))(F(6)); END;
DO I=1 TO 12; TM1(I)=TM1(I)+TM2(I); END;
PUT SKIP EDIT('  1+2 ')(A(6));
DO I=1 TO 12; PUT EDIT(TM1(I))(F(6)); END;
PUT SKIP(2) EDIT('  DECADE OF FIRST AND LAST RECORD :',
DE,' - ',DL)(A(40),X(1),F(1),A(3),F(1));
PUT SKIP EDIT('  TOTAL NUMBER OF RECORDS : ',SUM(T(*,*)))
(A(40),F(6));
K=0; DO I=1 TO 32; DO J=1 TO 28; IF T(I,J)¬=0 THEN K=K+1; END; END;
PUT SKIP EDIT('  NUMBER OF OCCUPIED GRIDS : ',K)
(A(40),F(6));

PUT SKIP(5);
ON ENDPAGE(SYSPRINT);
PUT SKIP EDIT('     ')(A); DO J=2 TO 28; PUT EDIT(J)(F(4)); END;
DO I=1 TO 32; PUT SKIP(2) EDIT(I)(F(2)); DO J=1 TO 28;
IF TH(I,J)=0 THEN PUT EDIT('    ')(A(4));
ELSE IF T(I,J)=0 THEN PUT EDIT('   .')(A(4));
ELSE PUT EDIT(T(I,J))(F(4));
END; END;

ON ENDPAGE(SYSPRINT) PUT PAGE;
PUT PAGE;
T=0; TM1=0; TM2=0; DE=8; DL=0; ISNV=ISNB;
IF ISW=0 THEN GO TO LABC;

PUT SKIP LIST('  END OF THE JOB.');
END KRT1;
```

Alphabetical index of the species

After each name the species number and, for the species, also the generic name is given. Subgenus names, infraspecific taxa, synonyms, and homonyms are printed in italics.